AMATEUR
ASTRONOMER'S
PHOTOGRAPHIC
LUNAR
ATLAS

AMATEUR ASTRONOMER'S PHOTOGRAPHIC

LUNAR ATLAS

Henry Hatfield

LUTTERWORTH PRESS
London

First published 1968

7188 1353 7

Printed in Great Britain
by Ebenezer Baylis and Son Ltd.
The Trinity Press, Worcester, and London.

EXPLANATORY NOTES

The atlas is divided into sixteen sections, each of which is made up of five or more photographic plates and a map. Each map is based primarily on the facing plate. Where detail is lacking on this primary plate, particularly near the Moon's limb, supplementary detail is provided from one or more of the other plates in a particular section.

The Maps—Overlaps—Scale—Map Grids

The maps and plates show what will be seen through the eyepiece of an astronomical telescope. Thus North is always at the bottom, and East is to the right. Each map overlaps its neighbours by a generous amount. No attempt has been made to adhere rigidly to the boundaries of the various numbered sections in the "Key to Maps and Plates"; indeed in some cases the boundaries of the plates have purposely been allowed to encroach into neighbouring sections. The Key Plate is intended to guide the reader into the right area; the "Index of Named Formations" lists every map on which a particular feature will be found.

The 'scale' of each map, and of the main plates which accompany it, has been adjusted so that the Moon's diameter is 25 inches (any variation in this diameter is stated in the relevant caption). Despite this the true scale (the relationship between the diameter of a crater on the Moon and its diameter in this atlas) of each map and plate varies from place to place; furthermore the North—South scale may well vary in a different way from the East—West scale. This is part of the very nature of the orthographic projection, in which the observer views a globe from a great distance. In order to give the reader some idea of the true scale from place to place, the diameters of five craters have been noted beneath each map. As the scale varies from plate to plate, so the size and shape of the various photographic images will also vary.

The grids on the various maps are intended to be used only for reference purposes. They bear no relationship to lunar latitude and longitude, or to any of the cartographic grids which are in use at the present time. If a formation appears on more than one map, then its grid references on each map will almost certainly vary. An unnamed and un-lettered formation on any map may be identified with certainty by quoting the map number and grid reference, e.g. Map **3** Square a4, and then making a small tracing of the square concerned, showing the formation.

Authority—Nomenclature

All the physical information given in this atlas, as well as the names or letters attaching to the various formations, has been accepted from *Named Lunar Formations* by Mary A. Blagg and K. Müller. The names and letters appearing on the various maps have been taken from "Vol. II Maps" of the same work. Although this atlas by Blagg and Müller was first published in 1935, it has seemed desirable to accept it in preference to more modern works, both to preserve the traditional names and letters, and to avoid the controversies which have sprung up concerning the re-naming and re-lettering of various formations. Some seven names, additional to those listed by Blagg and Müller, are indicated with an asterisk where they appear in the Index.

The Libration Keys

Beneath each main plate there is a small key, which shows the numbered area in which the plate lies, and the Moon's optical libration when the photograph was taken. The Moon does not present exactly the same face to the Earth all the time, but rocks gently back and forth in all directions, so that at one time or another an observer on the Earth will see about 60% of its surface. Referring to these libration keys, the reader should imagine that the Moon has rotated in the direction of the arrow by the amount indicated. Thus in Plate 1a the movement is 7·3° in a direction a little West of North, and therefore an area just East of the South Pole, which could not be seen if the Moon were in its mean position, has been exposed. If the blacked in 'square' on these keys lies in the same semicircle as the libration arrow, then the direction of the libration is fairly favourable for the area depicted on the plate; if the arrow passes through the square concerned then the direction is very favourable; an unfavourable libration will exist if the square concerned lies beyond the head of the libration arrow. The amount of libration can vary between nothing and a maximum of about 10°; 7° or 8° is considered to be good in most cases.

The Moon's physical libration and diurnal libration have been ignored, since their combined effect would not alter the aspect of the various photographs appreciably.

MAPS

and

PLATES

Key to Maps and Plates

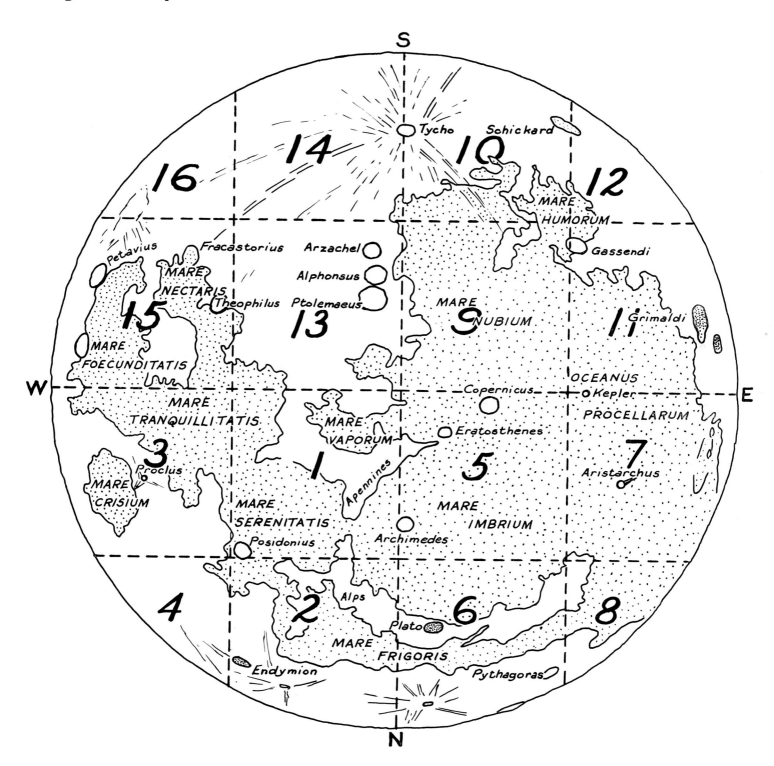

2210 G.M.T. 20.10.64.

Map 1

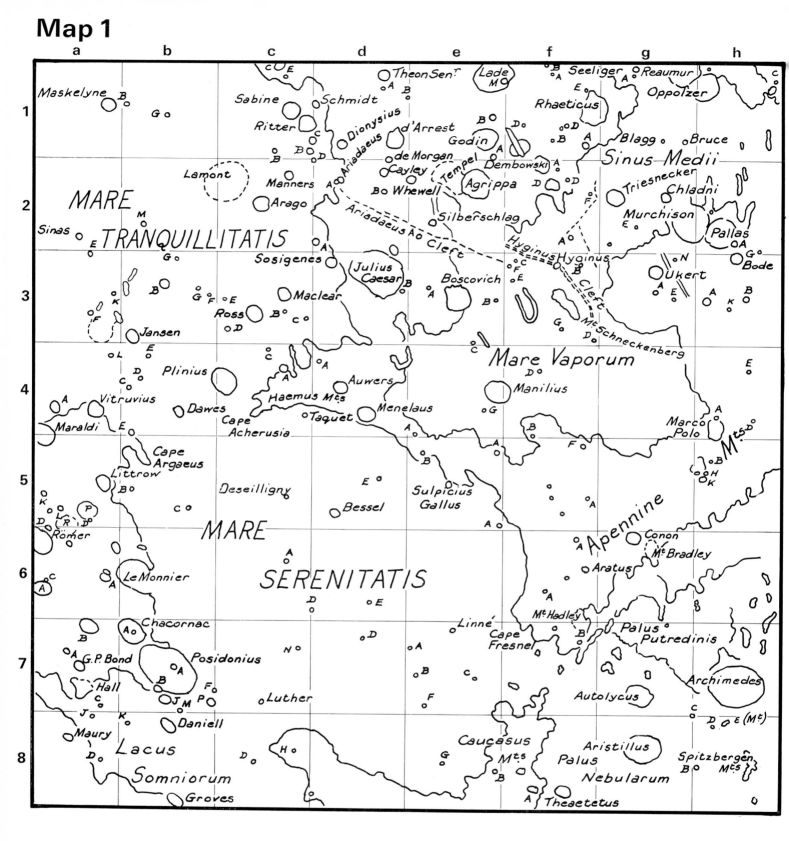

Crater Diameters

Archimedes	75 kms. (h7)
Posidonius	96 kms. (b7)
Manilius	36 kms. (e4)
Maskelyne	24 kms. (a1)
Bruce	9 kms. (g1)

This map has been prepared from Plate 1a. Plates 1b, 1c, and 1d show the same area under different lighting. Plate 1e shows larger scale photographs of the Triesnecker (g2) and Linné (e7) areas.

Plate 1a

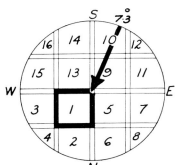

2046 G.M.T. 28.4.66. Moon's age 7·9 days. Diameter 25 inches. The Apennine Mts. (g5) rise to 16,000 feet in places.

Plate 1b

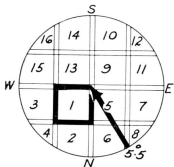

0215 G.M.T. 6.8.66. Moon's age 18·9 days. Diameter 25 inches. Note the ridges on the mare to the SE of Posidonius (b7), and the domes near Arago. These are shown on a larger scale on Plate 3f.

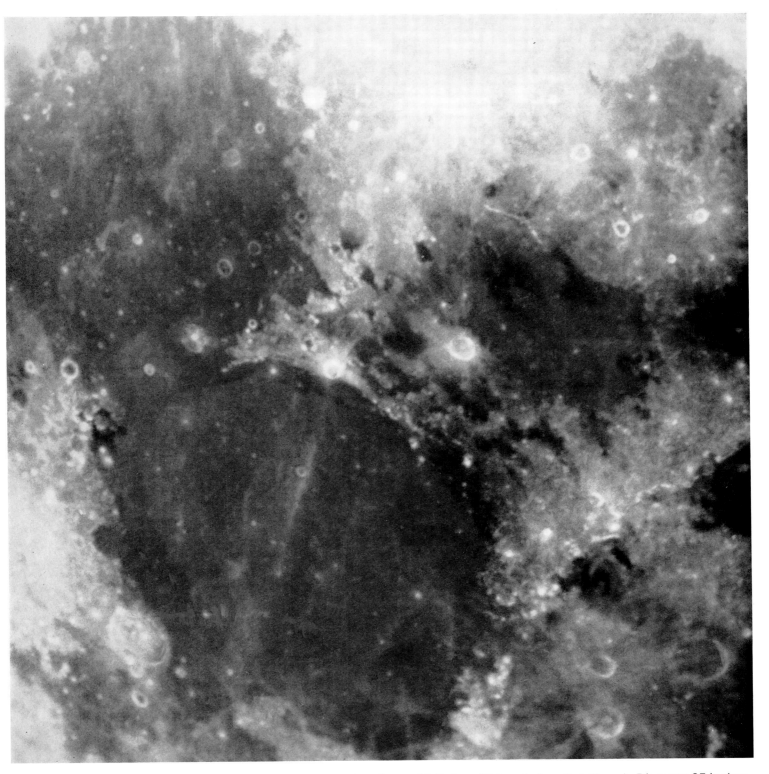

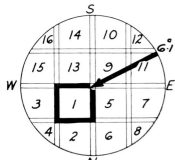

2137 G.M.T. 25.12.66. Moon's age 13·8 days (2 days before Full Moon). Diameter 25 inches. Compare this with Plate 1a, which shows almost the same area.

Plate 1d

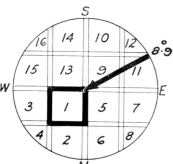

1835 G.M.T. 18.3.67. Moon's age 7·6 days. Diameter 25 inches. Although the Moon's age here is almost the same as in Plate 1a, the aspect is different, because of the different libration.

Linné, Bessel and Sulpicius Gallus in the Mare Serenitatis. 2017 G.M.T. 16.5.67. Moon's age 7·2 days. Diameter 37 inches approx. Linné lies in Map **1** e7.

Triesnecker, Hyginus and their systems of clefts. 2016 G.M.T. 16.5.67. Moon's age 7·2 days. Diameter 37 inches approx. Triesnecker lies in Map **1** g2.

Map 2

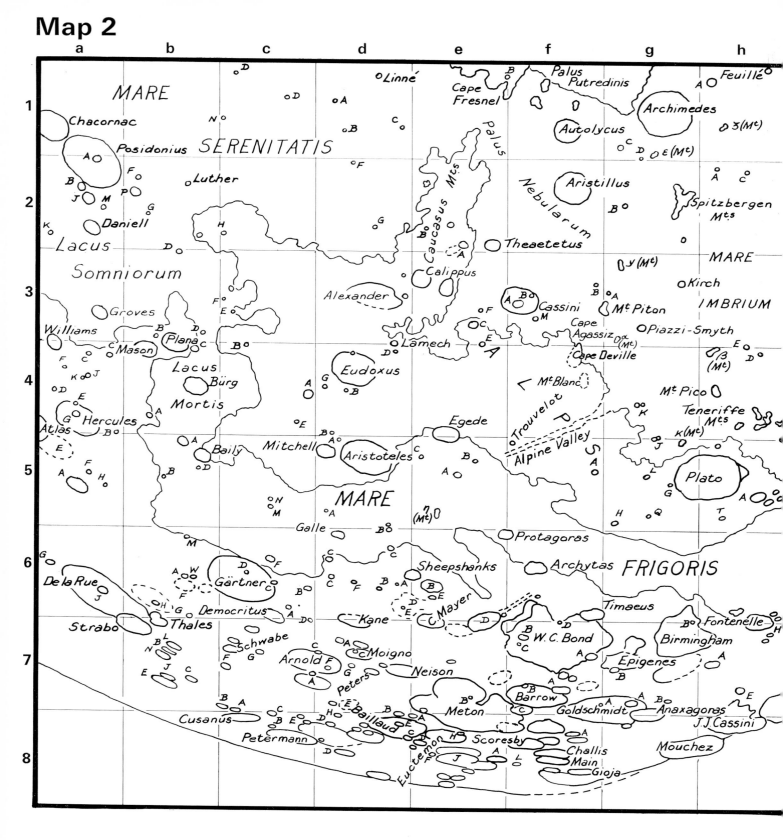

Crater Diameters

Autolycus	36 kms. (f1)
Daniell	24 kms. (a2)
Egede	49 kms. (e4)
Strabo	47 kms. (b7)
Gioja	35 kms. (f8)

This map has been prepared from Plate 2a. Plates 2b, 2c and 2d show the same area under different lighting. Plate 2e shows larger scale photograph of the Alpine Valley (f4) area.

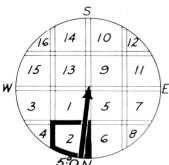

1814 G.M.T. 22.11.66. Moon's age 10·2 days. Diameter 25 inches. This is a very favourable northerly libration, which is not likely to be seen very often. Note that Plato (h5) is very nearly circular. The Alps are about 12,000 feet high in places.

Plate 2b

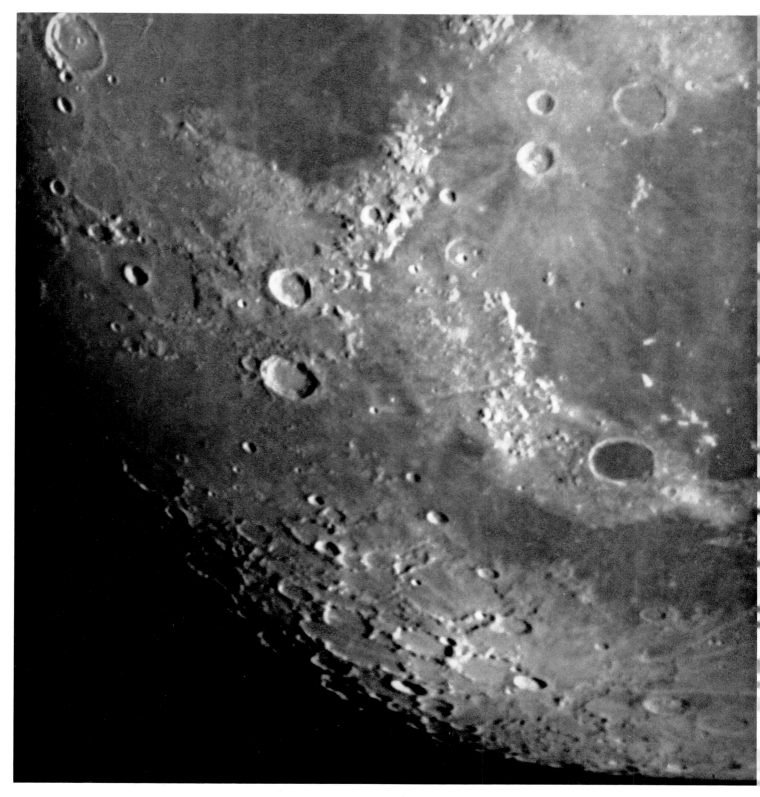

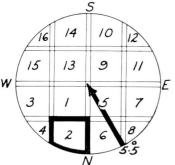

0211 G.M.T. 6.8.66. Moon's age 18·9 days. Diameter 25 inches. Compare this libration with that of Plate 2c.

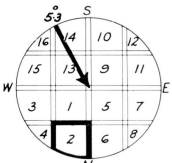

2103 G.M.T. 29.5.66. Moon's age 9·4 days. Diameter 25 inches. This is a relatively bad libration for this area. Note how much closer Plato (h5) is to the limb here than in Plates 2a and 2b.

Plate 2d

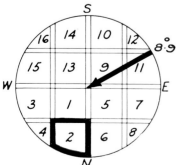

1835 G.M.T. 18.3.67. Moon's age 7·6 days. Diameter 25 inches. Note how Mt. Piton (g3) catches the early morning sunlight. The Caucasus Mts. (e2) rise to about 20,000 feet.

Plate 2e

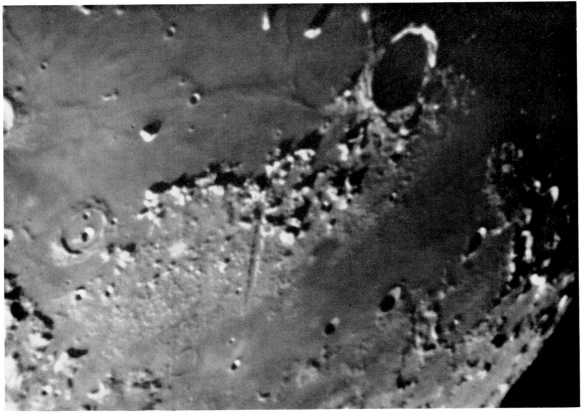

Cassini, the Alpine Valley and Plato. 1925 G.M.T. 19.3.67. Moon's age 8·6 days. Diameter 36 inches approx. Note the large "ghost" crater ring just South of Plato. This is sometimes called "Ancient Newton".

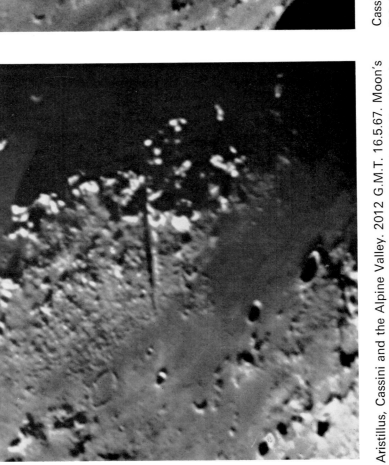

Aristillus, Cassini and the Alpine Valley. 2012 G.M.T. 16.5.67. Moon's age 7·2 days. Diameter 37 inches approx. The Alpine Valley lies in Map **2** f4.

Map 3

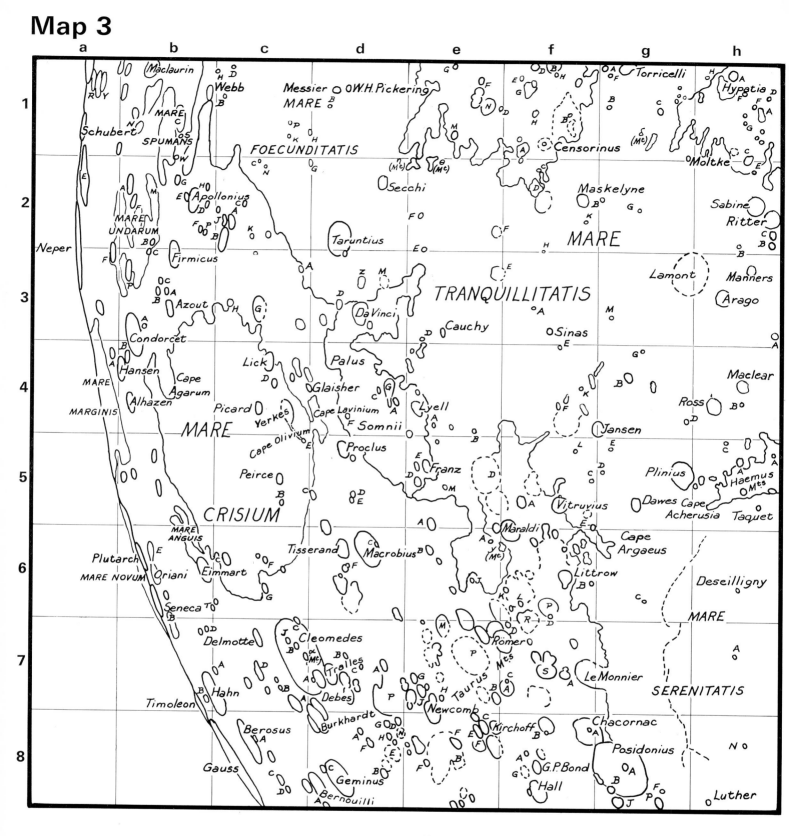

Crater Diameters

Luther	8½ kms.	(h8)
Berosus	61 kms.	(c8)
Lyell	43½ kms.	(e4)
Sabine	31 kms.	(h2)
Maclaurin	45 kms.	(b1)

This map has been prepared from Plates 3a and 3d. Plates 3b, 3c and 3e show the same area under different lighting. Plate 3f shows larger scale photographs of the area near Arago (h3), and the Wrinkle Ridges east of Le Monnier (g7). Plate 3g shows Messier and Pickering (d1).

Note: The position of Timoleon (b7) has been accepted from Blagg and Müller. Many people consider that the correct "Timoleon" is the much larger crater which lies to the West of this one.

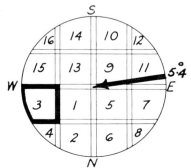

1757 G.M.T. 16.2.67. Moon's age 7·3 days. Diameter 25 inches. Note the rays round Proclus (d5). There are several "domes" North and East of Arago (h3).

Plate 3b

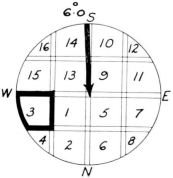

2228 G.M.T. 8.1.66. Moon's age 17·0 days. Diameter 25 inches. Compare the shape of the Mare Crisium (c5) here with that in Plate 3a.

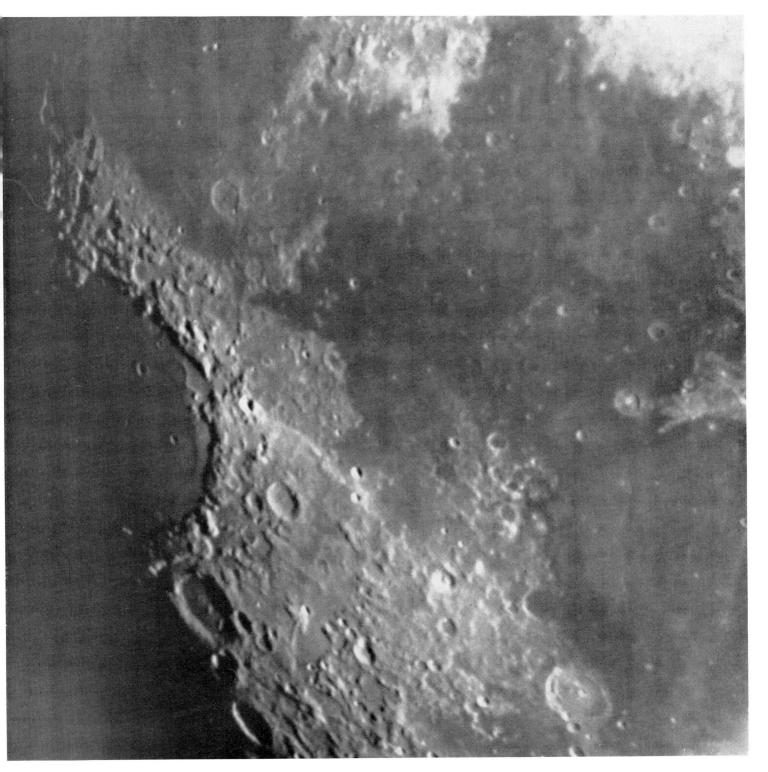

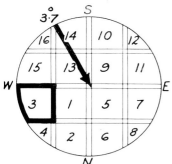

2336 G.M.T. 26.2.67. Moon's age 17·6 days. Diameter 25 inches. Compare the Mare Tranquillitatis (f3) here with the same area in Plate 3a.

Plate 3d

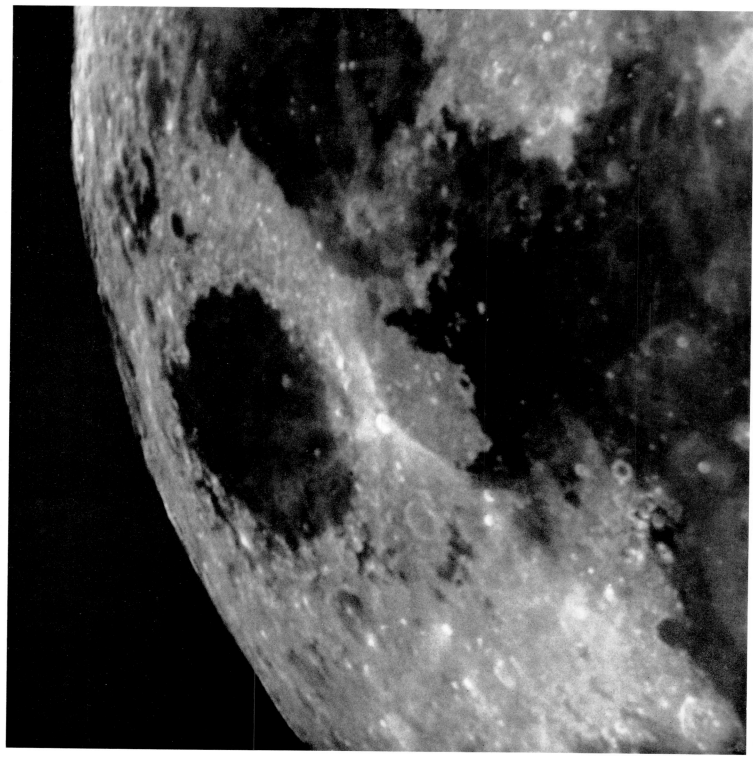

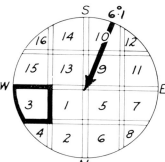

2353 G.M.T. 24.2.67. Moon's age 15·6 days. Diameter 25 inches. This was taken 6 hours after Full Moon.

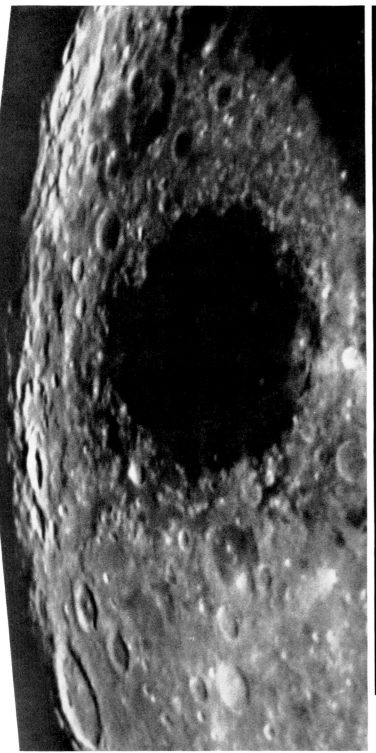

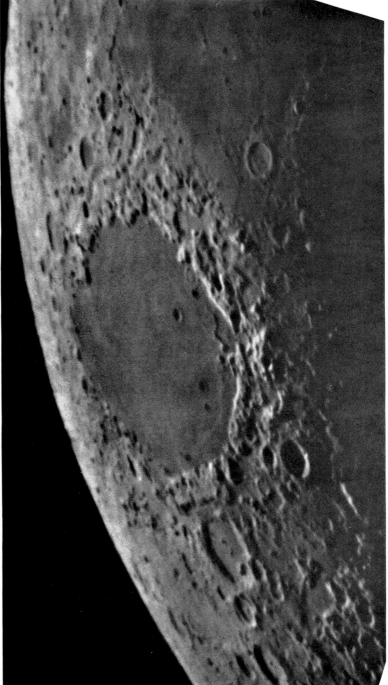

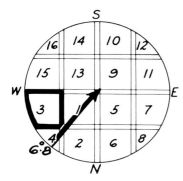

2304 G.M.T. 20.8.67. Moon's age 14·9 days. Diameter 25 inches. Gauss (c8) is near the limb and Neper (a2) is on the limb.

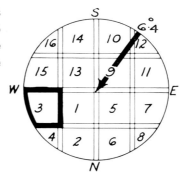

2034 G.M.T. 23.5.66. Moon's age 3·4 days. Diameter 25 inches. Compare this with Plate 3b and with its neighbour here.

Plate 3f

Arago, its "Domes" and the Ariadaeus Cleft. 2000 G.M.T. 15.5.67. Moon's age 6·2 days. Diameter 37 inches approx. Arago lies in Map **3** h3, and Map **1** c2. The domes lie about its own diameter North and East of it.

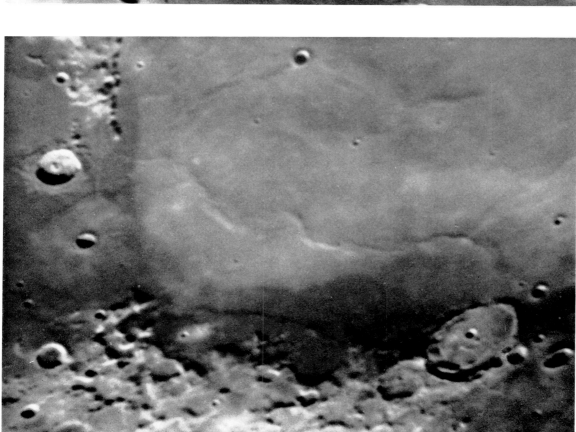

Plinius, Posidonius, Bessel and the "Wrinkle Ridges" in Mare Serenitatis. 2015 G.M.T. 15.5.67. Moon's age 6·2 days. Diameter 37 inches approx. Plinius lies in Map **3** g5.

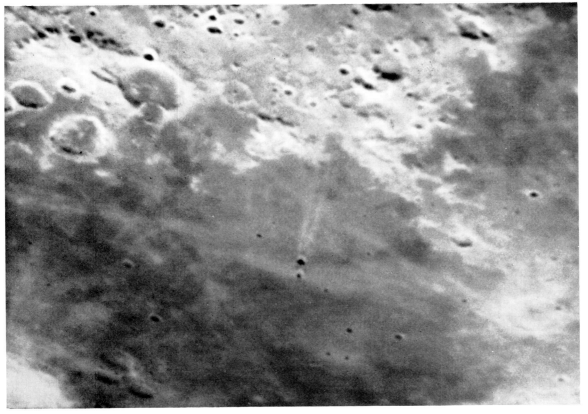

Messier and Pickering. 2008 G.M.T. 15.5.67. Moon's age 6·2 days. Diameter 37 inches approx.

Messier and Pickering. 2002 G.M.T. 13.5.67. Moon's age 4·2 days. Diameter 37 inches approx.

Note how these two craters change their appearance in two days. They lie in Map **3** d1.

Map 4

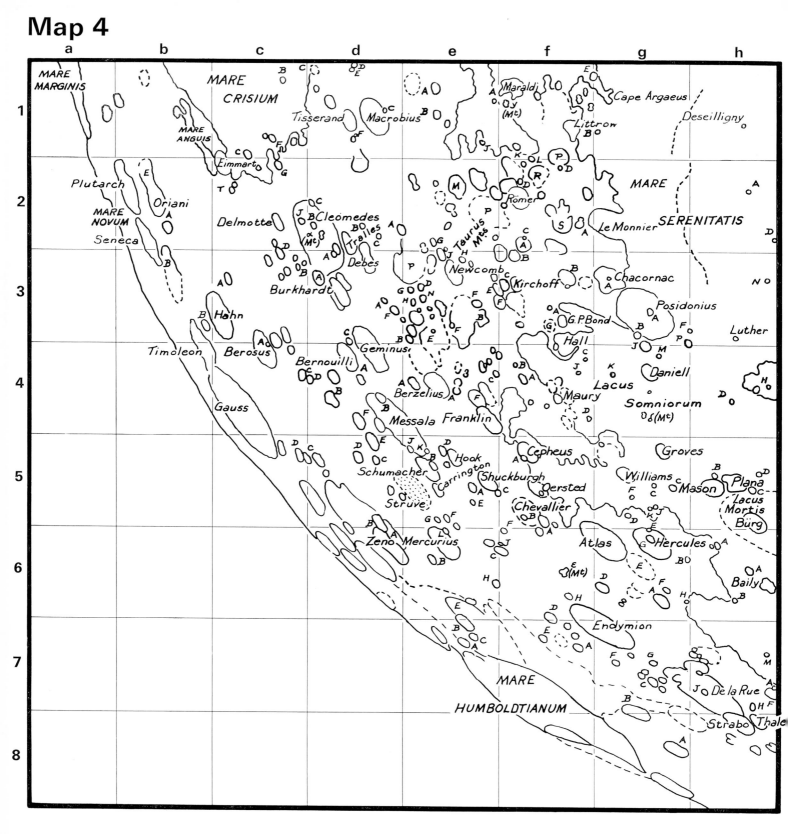

MARE MARGINIS

MARE CRISIUM

MARE ANGUIS

Plutarch

Oriani

MARE NOVUM

Seneca

Tisserand

Eimmart

Delmotte

Cleomedes

Trailes

Debes

Burkhardt

Hahn

Berosus

Timoleon

Bernouilli

Geminus

Gauss

Berzelius

Messala

Franklin

Schumacher

Carrington

Struve

Zeno

Mercurius

Macrobius

Taurus Mts.

Newcomb

Kirchoff

Maraldi

(Mt)

Littrow

Römer

Le Monnier

Chacornac

Posidonius

G.P.Bond

Hall

Maury

Cepheus

Shuckburgh

Oersted

Chevallier

Cape Argaeus

Deseilligny

MARE SERENITATIS

Luther

Daniell

Lacus Somniorum

(Mt)

Groves

Williams

Mason

Plana

Lacus Mortis

Bürg

Atlas

Hercules

Baily

Hook

Endymion

MARE HUMBOLDTIANUM

De la Rue

Strabo

Thale

Crater Diameters

Endymion	117 kms. (g7)
Mason	31 kms. (h5)
Gauss	136 kms. (c4)
Römer	36½ kms. (f2)
Tisserand	33 kms. (d1)

This map has been prepared from Plates 4a and 4e. Plates 4b, 4c and 4d show the same area under different lighting.

Note: The position of Timoleon (b4) has been accepted from Blagg and Müller. Many people consider that the correct "Timoleon" is the much larger crater which lies to the West of this one and which is right on the terminator on Plate 4e.

1757 G.M.T. 16.2.67. Moon's age 7·3 days. Diameter 25 inches.

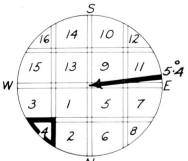

Plate 4b

2034 G.M.T. 23.5.66. Moon's age 3·4 days. Diameter 25 inches.

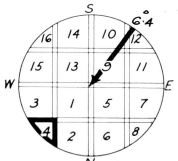

2127 G.M.T. 28.11.66. Moon's age 16·3 days. Diameter 25 inches.

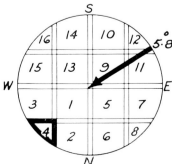

Plate 4d

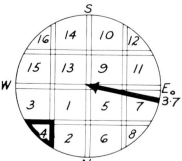

2248 G.M.T. 29.10.66. Moon's age 15·7 days. Diameter 25 inches. Compare this with Plate 4e. The area is the same in each case but the libration is very different.

Plate 4e

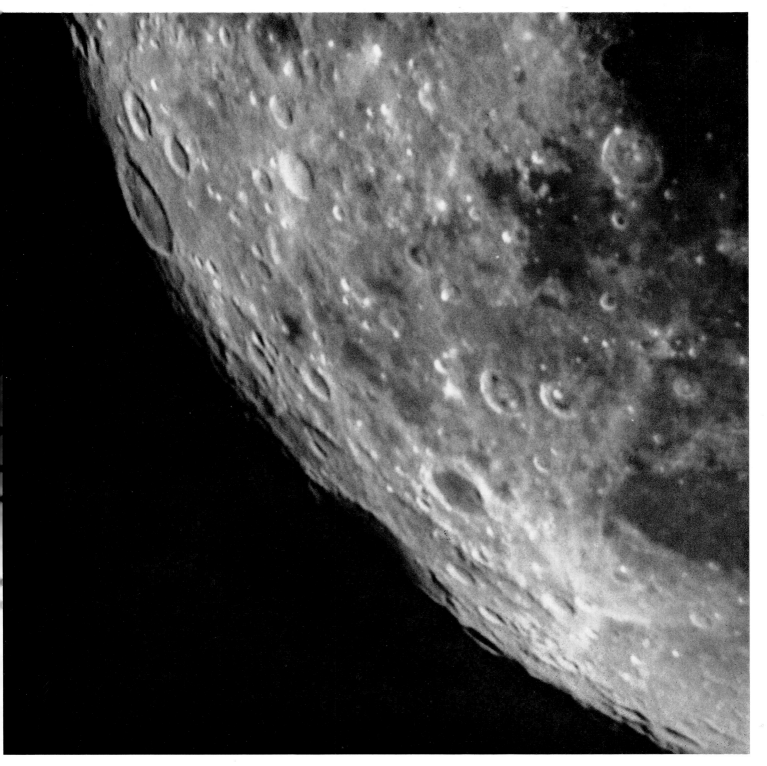

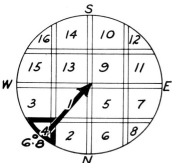

2304 G.M.T. 20.8.67. Moon's age 14·9 days. Diameter 25 inches. This is quite a good libration for this area. The Mare Humboldtianum (f7) and Gauss (c4) do not often appear like this.

Map 5

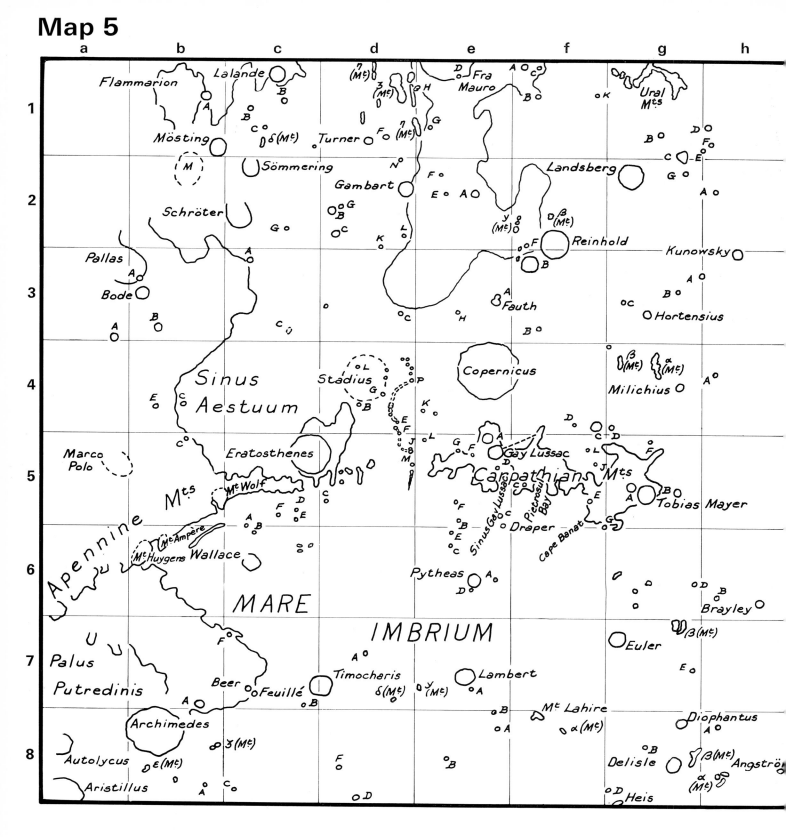

Crater Diameters

Delisle	22½ kms.	(g8)
Beer	10½ kms.	(c7)
Copernicus	97 kms.	(e4)
Landsberg	42 kms.	(g2)
Mösting	26 kms.	(b1)

This map has been prepared from Plate 5a. Plates 5b, 5c and 5d show th[e] same area under different lighting. Plate 5e shows larger scale photograph[s] of the area around Copernicus (e4) and of the "Domes" in the vicinity [of] Milichius (g4).

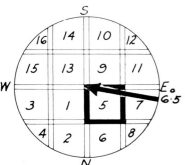

0315 G.M.T. 9.8.66. Moon's age 23·8 days. Diameter 25 inches. Note the ridges in the Mare Imbrium and the Sinus Aestuum (c4). The Carpathian Mts. rise to about 7,000 feet in places.

Plate 5b

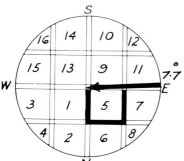

1947 G.M.T. 31.1.66. Moon's age 10·1 days. Diameter 25 inches. Compare this with Plate 5c.

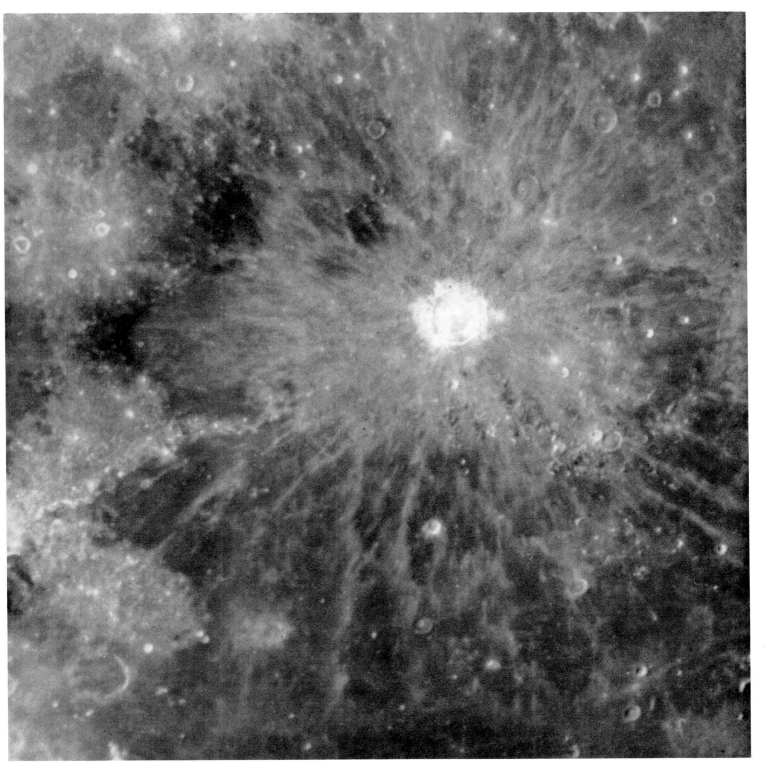

2135 G.M.T. 25.12.66. Moon's age 13·8 days. Diameter 25 inches. This photograph shows almost exactly the same area as Plate 5b.

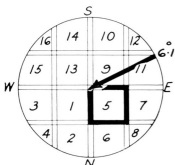

Plate 5d

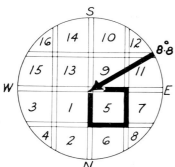

1801 G.M.T. 19.2.67. Moon's age 10·3 days. Diameter 25 inches. This photograph extends further to the West than the others in this group.

Euler to Hortensius, with the "Domes" near Milichius. 1908 G.M.T. 21.3.67. Moon's age 10·6 days. Diameter 36 inches approx. Milichius lies in Map **5** g4; there are several domes near it here, mostly to the North-East.

The Copernicus area soon after sunrise. 2041 G.M.T. 20.3.67. Moon's age 9·7 days. Diameter 36 inches approx. Draper (e6) is $8\frac{1}{2}$ kms. in diameter.

Map 6

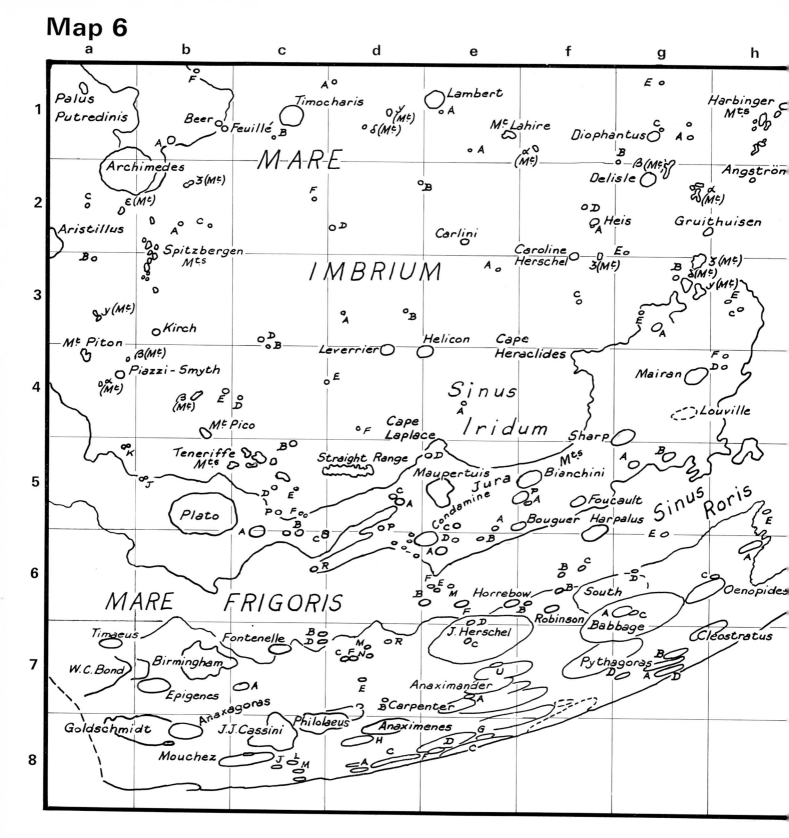

Crater Diameters

Cleostratus	70 kms.	(h7)
Epigenes	52 kms.	(b7)
Leverrier	24½ kms.	(d4)
Angström	10½ kms.	(h2)
Timocharis	35 kms.	(c1)

This map has been prepared from Plate 6a. Plates 6b, 6c, and 6d show th
same area under different lighting. Plate 6e shows larger scale photograph
of the Straight Range (d5) and surrounding country, and of the Sinu
Iridum (e4) soon after Sunrise.

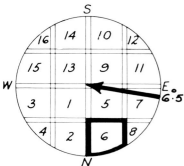

0315 G.M.T. 9.8.66. Moon's age 23·8 days. Diameter 25 inches. Note the "ghost" craters and low ridges near the Western border of the Mare Imbrium. Mt. Pico (b4) is about 8,000 feet high; it is not nearly so steep as it looks.

Plate 6b

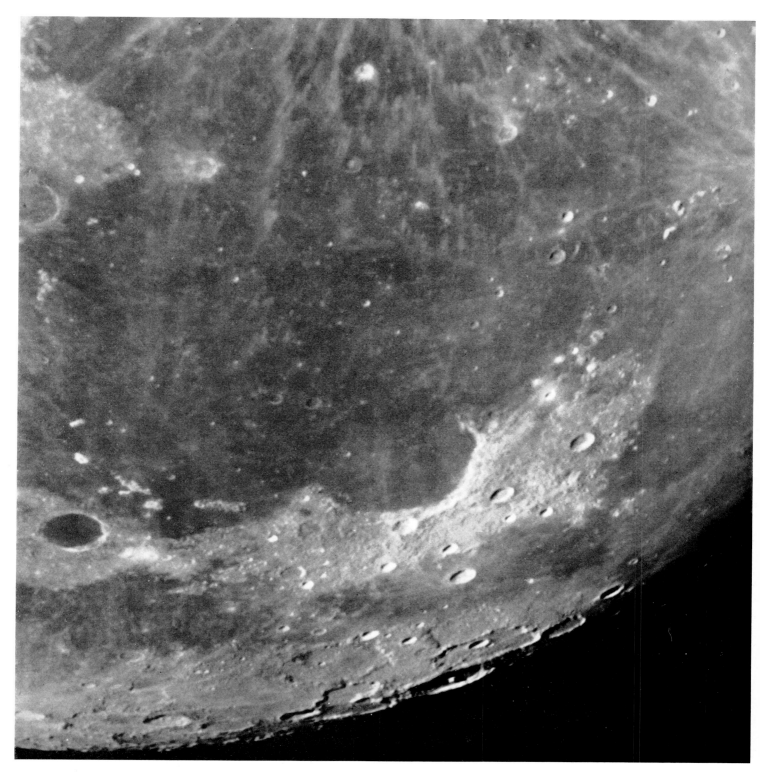

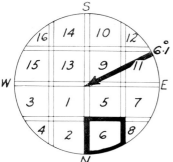

2040 G.M.T. 25.12.66. Moon's age 13·8 days. Diameter 25 inches. This was two days before Full Moon. Compare the shape of Plato (b5) here with that in Plate 6a. The large crater with a central mountain right on the terminator is Pythagoras (f7).

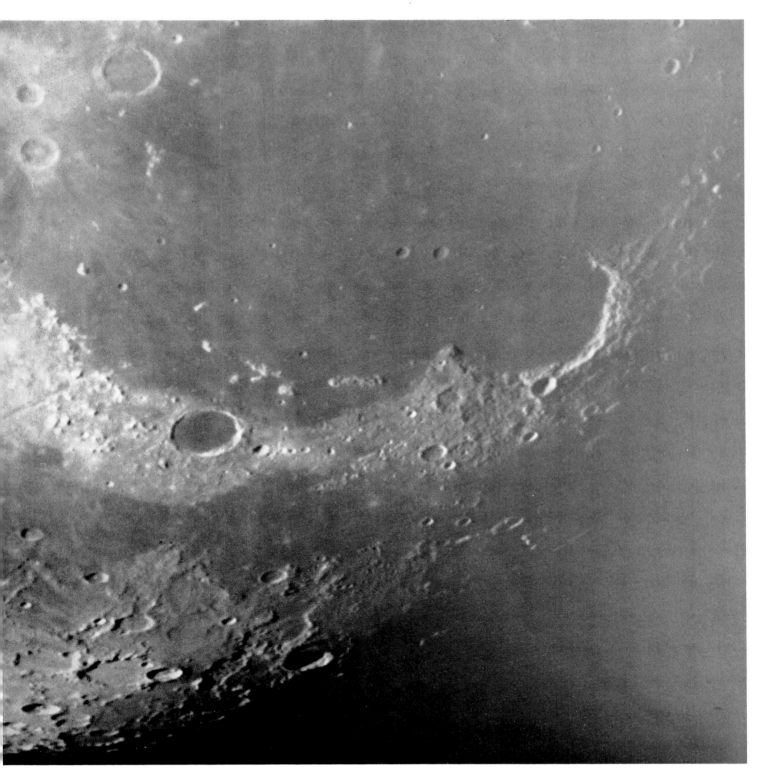

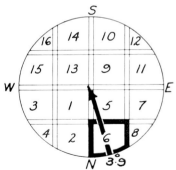

2142 G.M.T. 23.11.66. Moon's age 11·3 days. Diameter 25 inches. This is quite a good libration for the north polar regions. The bottom left-hand (NW) part of this photograph extends beyond Map 6; it is shown on Map 2.

Plate 6d

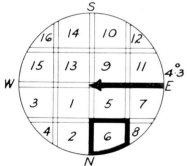

2232 G.M.T. 23.12.66. Moon's age 11·8 days. Diameter 25 inches. Compare this with Plate 6c, and note how the different librations cause the shapes of the craters to alter.

The Sinus Iridum soon after sunrise. 1923 G.M.T. 21.3.67. Moon's age 10·6 days. Diameter 36 inches approx. Sinus Iridum lies in Map **6** e4.

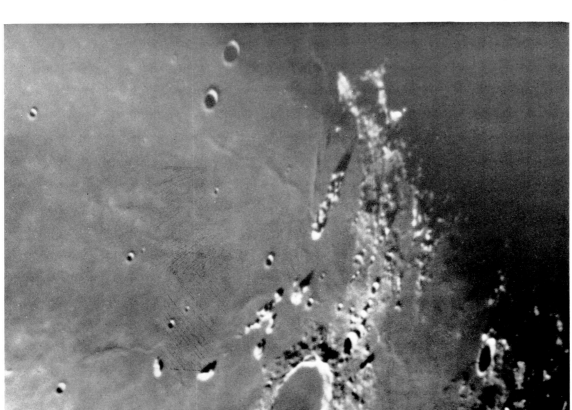

The Straight Range and surrounding country. 2029 G.M.T. 20.3.67. Moon's age 9·7 days. Diameter 36 inches approx. The Straight Range, which lies in Map **6** d5, is a typical isolated mountain range. Its highest peaks are about 6,000 feet high.

Map 7

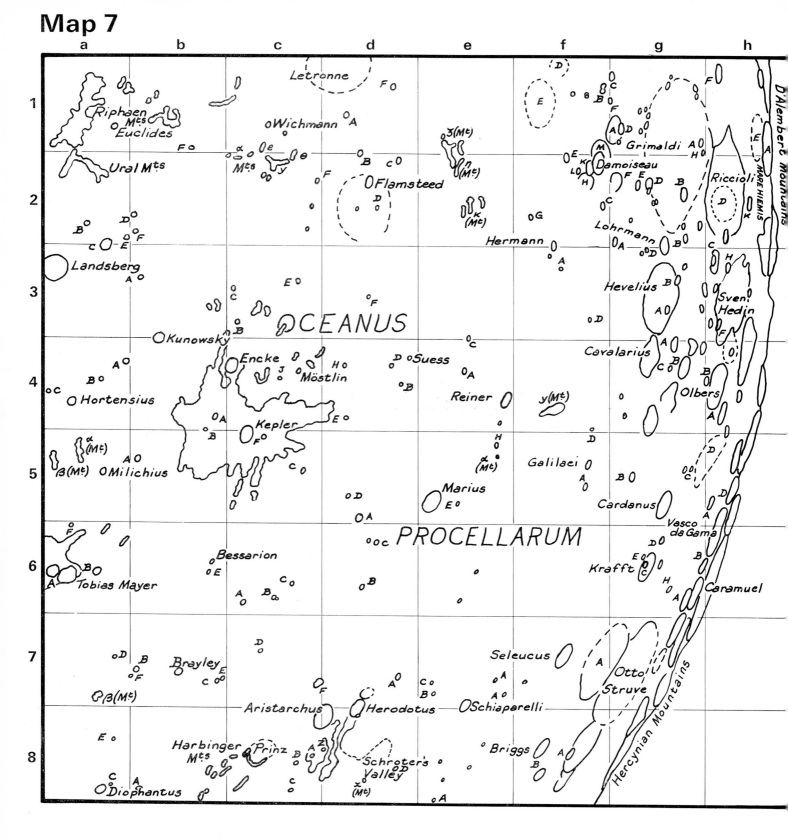

Crater Diameters

Briggs	38 kms. (f8)
Diophantus	17½ kms. (a8)
Marius	42 kms. (e5)
Hevelius	122 kms. (g3)
Euclides	12 kms. (a1)

This map has been prepared from Plates 7a and 7e. Plates 7b, 7c and 7 show the same area under different lighting. On Plate 7e there is an inse of the large limb crater Caramuel (h6), which is only exposed to view whe the libration is extremely favourable.

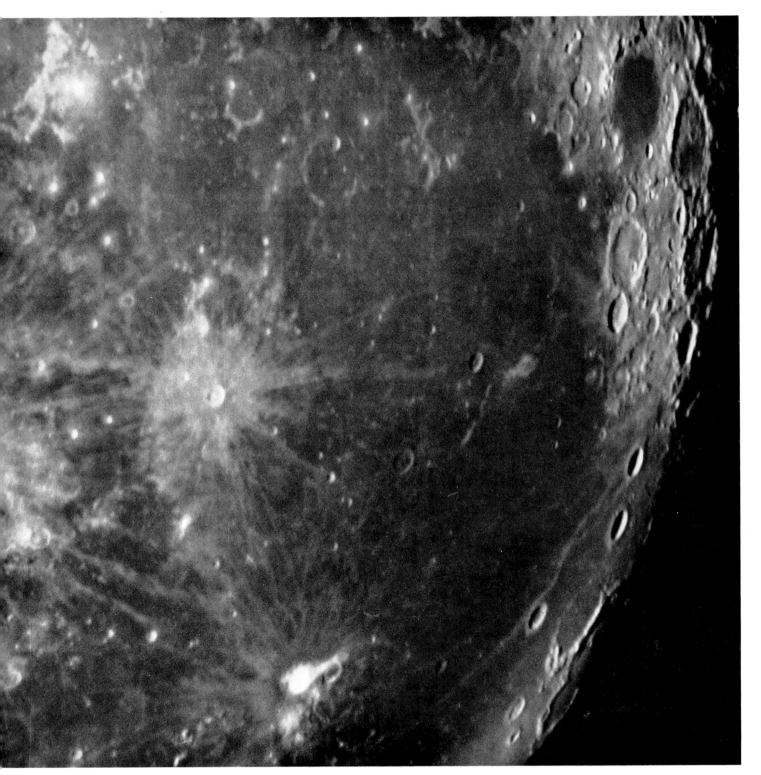

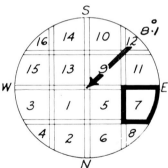

2310 G.M.T. 24.1.67. Moon's age 14·2 days. Diameter 25 inches. A comparatively favourable libration here has brought the "limb craters" into view much earlier than usual. Compare the dark floor of Grimaldi (g1) with the extreme brightness of Aristarchus (c8).

Plate 7b

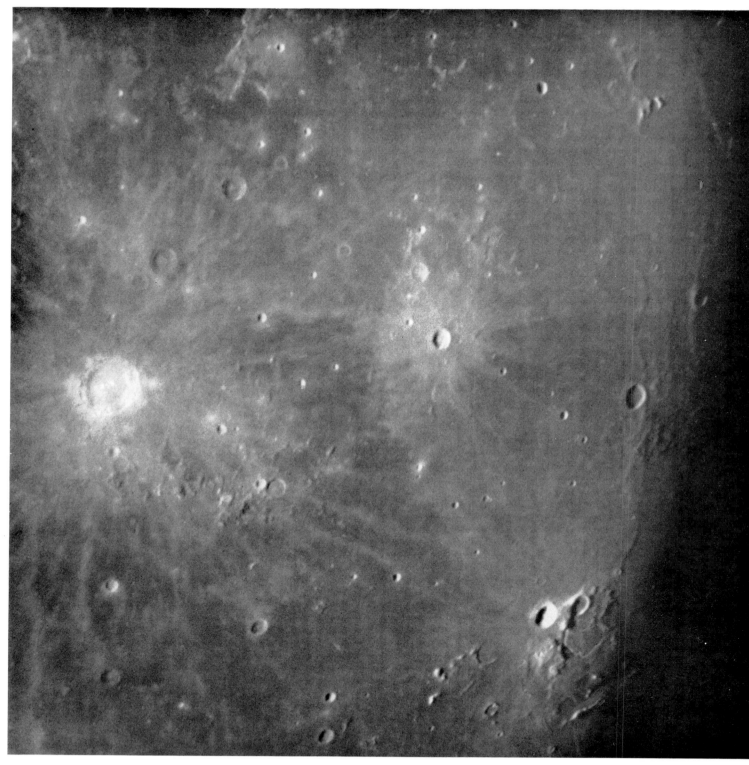

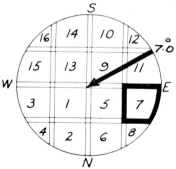

1944 G.M.T. 2.2.66. Moon's age 12·1 days. Diameter 25 inches. This photograph extends further to the West than the others in this set. Note Schröter's Valley (d8) and the low ridges near the terminator.

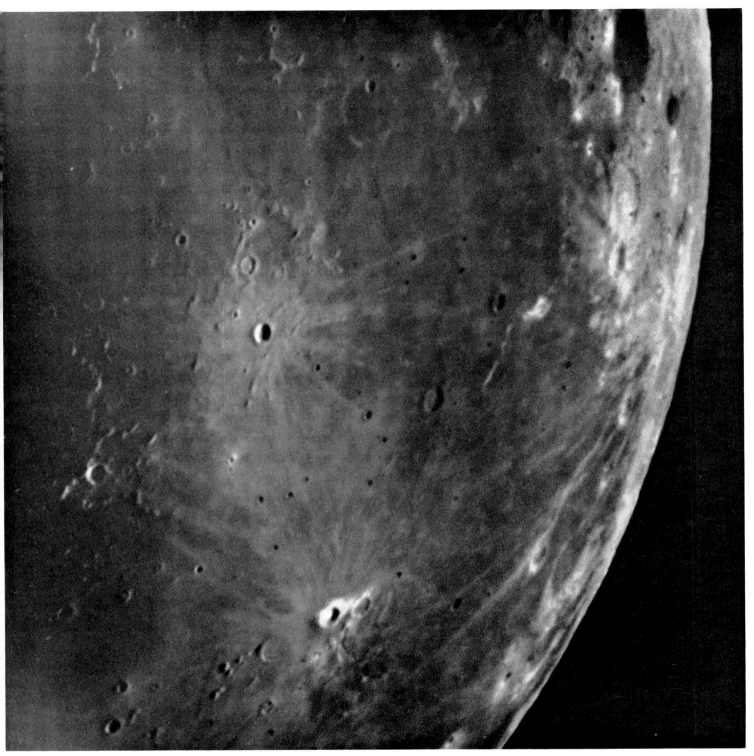

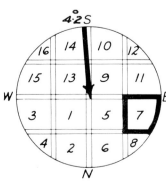

0655 G.M.T. 7.12.66. Moon's age 24·6 days. Diameter 25 inches. Compare this with Plate 7a, particularly near the limb.

Plate 7d

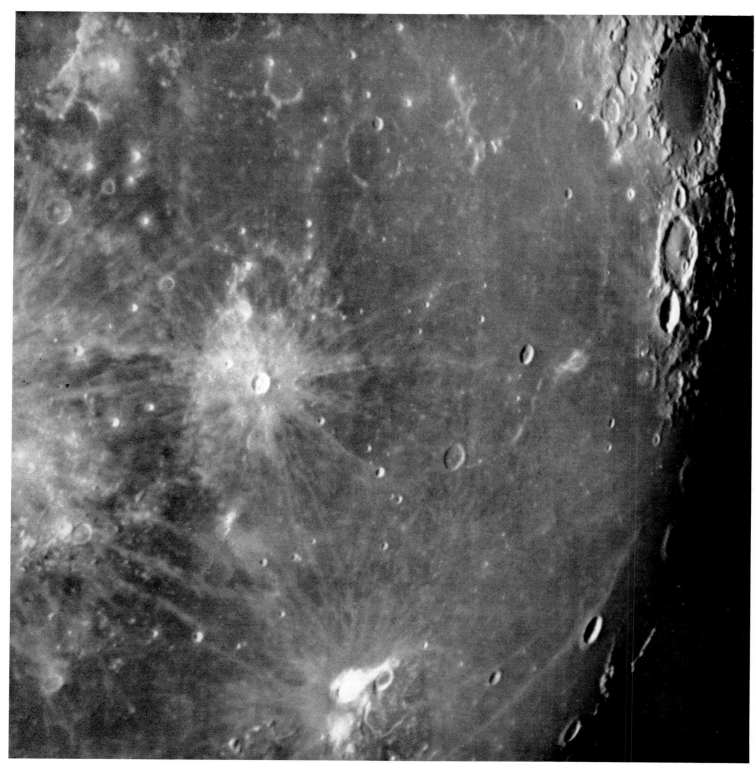

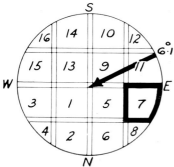

2036 G.M.T. 25.12.66. Moon's age 13·8 days. Diameter 25 inches. Note the "St. Andrew's Cross" marking on the Eastern wall of Grimaldi (g1).

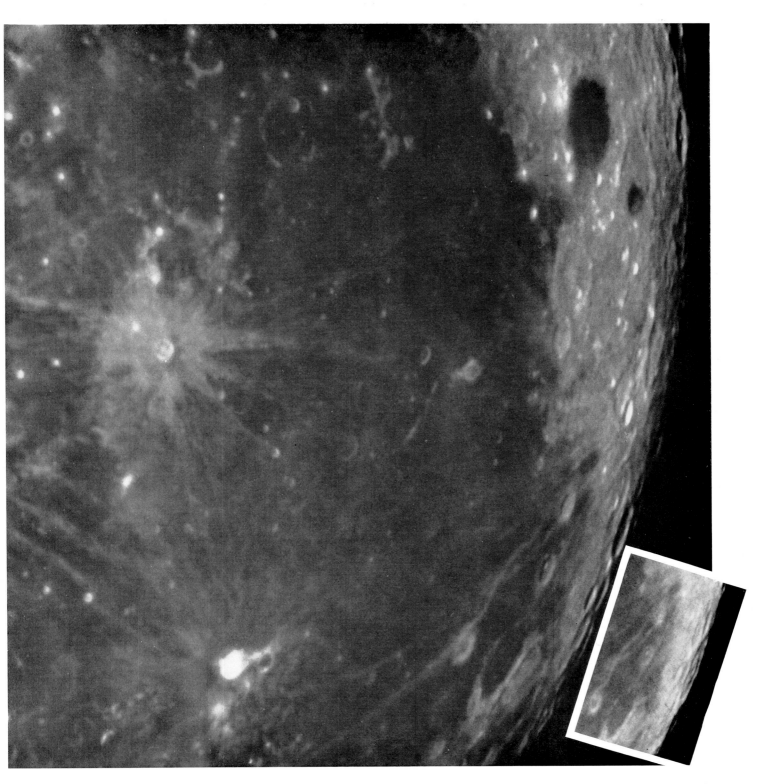

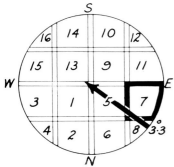

2219 G.M.T. 28.10.66. Moon's age 14·7 days. Diameter 25 inches. This was taken 12 hours before Full Moon. Caramuel (h6) is just beyond the terminator. Vasco da Gama (h6) is showing plainly. The insert shows Caramuel, taken at 2132 G.M.T. 8.11.65, with the author's six-inch reflector. This crater may be well seen on only one or two nights in the average year.

Map 8

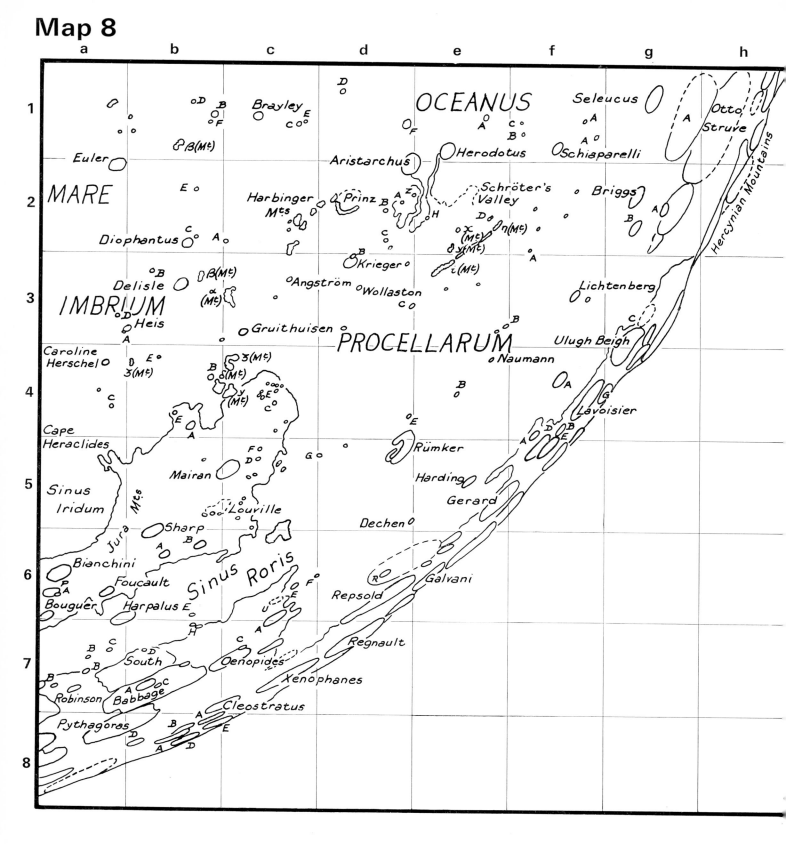

Crater Diameters

Pythagoras	113 kms. (a8)
Gerard	87 kms. (e5)
Seleucus	45 kms. (g1)
Wollaston	10½ kms. (d3)
Euler	24½ kms. (a2)

This map has been prepared from Plates 8a and 8c. The extreme NW corner comes from Plate 6a. Plates 8b and 8d show the same area under different lighting. Plate 8e shows larger scale photographs of the Schröter's Valley area (e2), Rümker (d5), and the Bands in Aristarchus (d2).

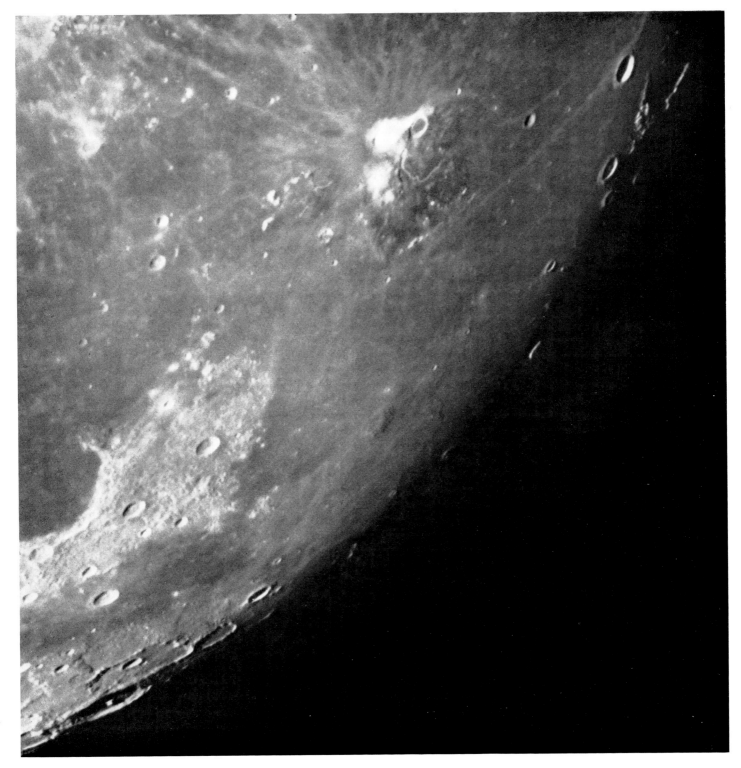

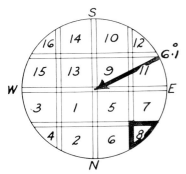

2040 G.M.T. 25.12.66. Moon's age 13·8 days. Diameter 25 inches. Pythagoras (a8) is on the terminator near the bottom. Rümker (d5) looks more like a mound than a crater.

Plate 8b

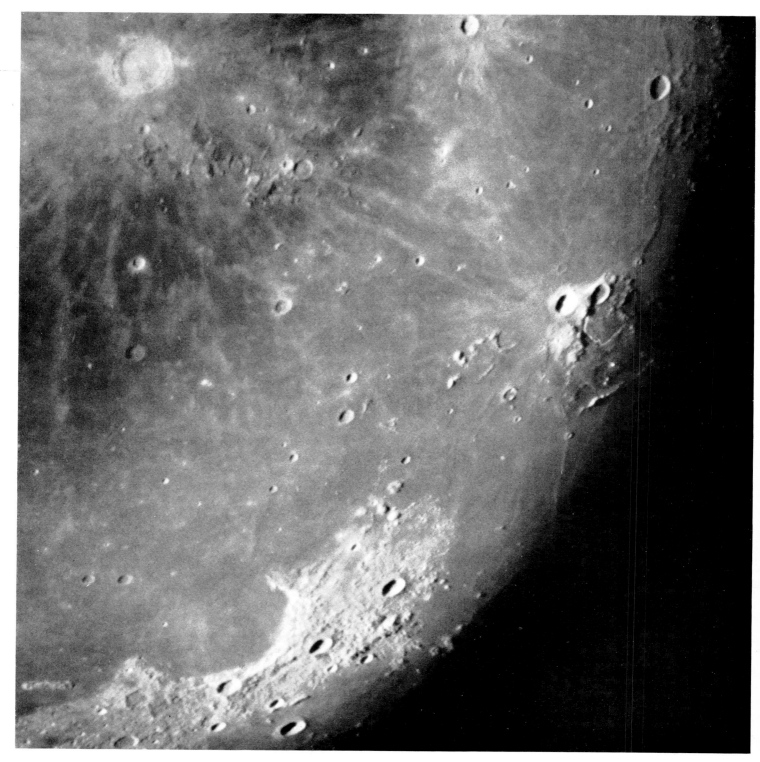

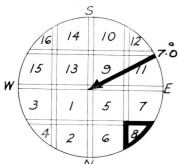

1944 G.M.T. 2.2.66. Moon's age 12·1 days. Diameter 25 inches. The Jura Mts. (a6) rise to about 20,000 feet. The Harbinger Mts. (c2) are about 8,000 feet high.

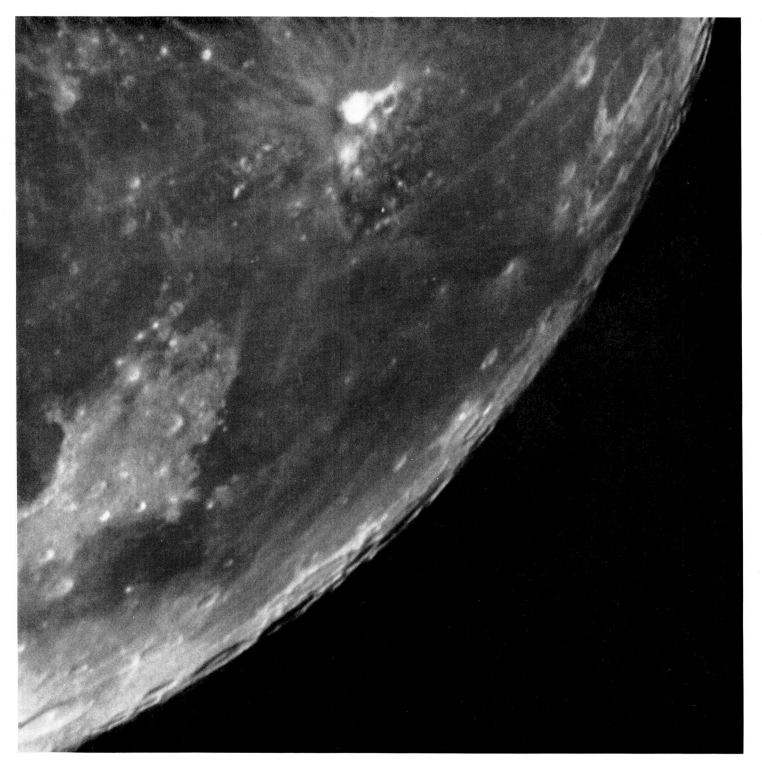

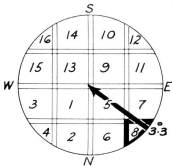

2219 G.M.T. 28.10.66. Moon's age 14·7 days. Diameter 25 inches. This was taken 12 hours before Full Moon. The libration is favourable; the craters on the terminator (limb) will not often show up like this.

Plate 8d

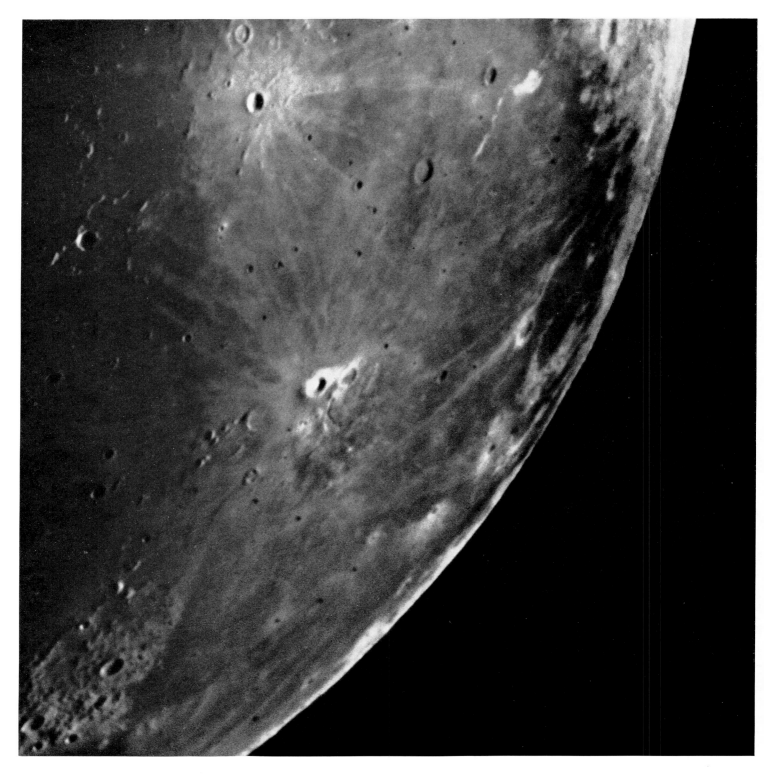

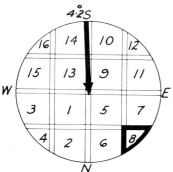

0655 G.M.T. 7.12.66. Moon's age 24·6 days. Diameter 25 inches. This is not a favourable libration. Many of the craters shown on Plate 8c are out of sight here, beyond the limb.

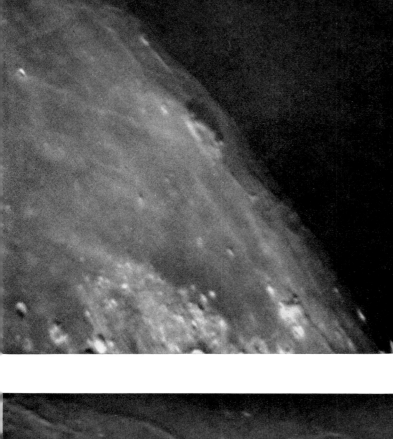

Rümker. 1944 G.M.T. 23.3.67. Moon's age 12·6 days. Diameter 35 inches approx. Rümker lies in Map **8** d5.

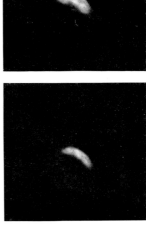

2110 G.M.T. 20.5.67. Moon's age 11·3 days. Band's in Aristarchus (d2). All these photographs had exposures much less than normal, so that Aristarchus was properly exposed; compare them with the Plate to the left. All are to the same scale, the Moon's diameter being 37 inches approx.

1952 G.M.T. 21.4.67. Moon's age 11·9 days.

2150 G.M.T. 23.4.67. Moon's age 14·0 days.

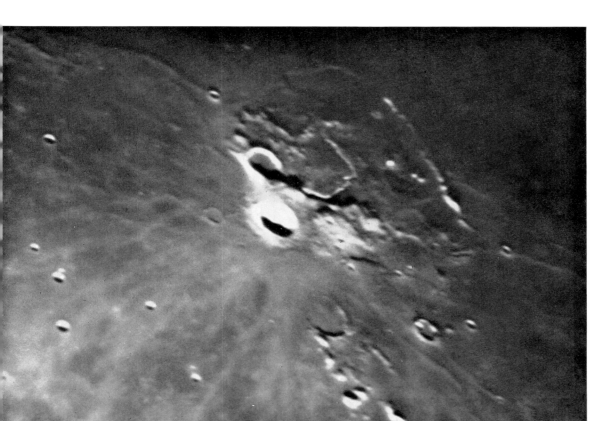

Schröter's Valley, Aristarchus and Herodotus. 2033 G.M.T. 21.4.67. Moon's age 12·0 days. Diameter 37 inches approx. Schröter's Valley lies in Map **8** e2.

Map 9

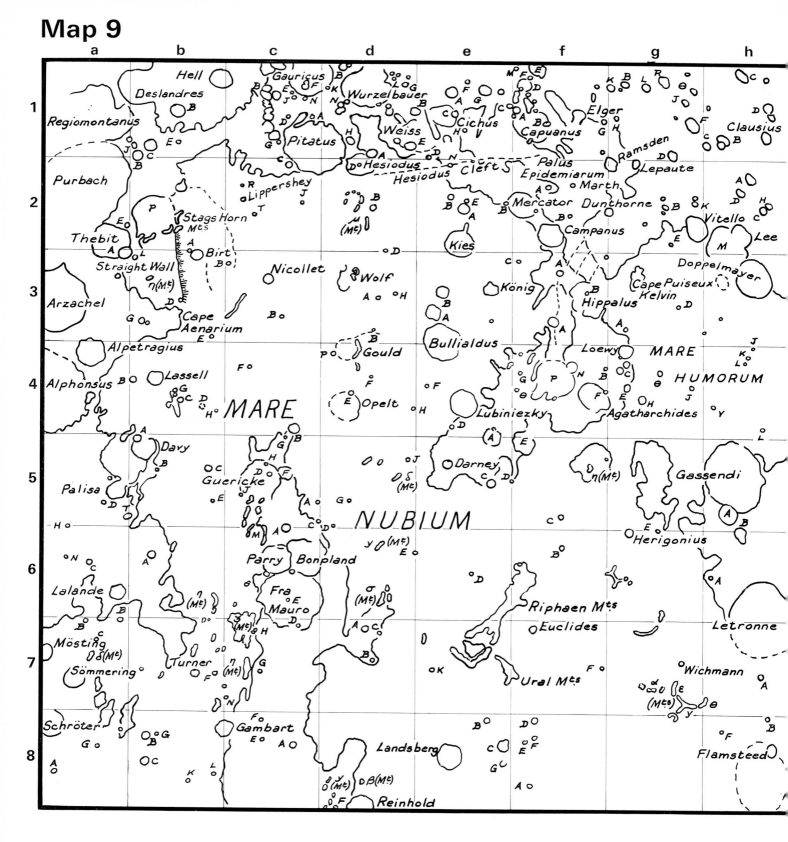

Crater Diameters

Flamsteed	19 kms. (h8)
Gambart	26 kms. (c8)
Bullialdus	50 kms. (e3)
Lepaute	14 kms. (g2)
Hell	31 kms. (b1)

This map has been prepared from Plate 9a. Plates 9b, 9c, 9d and 9e show the same area under different lighting. Plate 9f shows larger scale photographs of the Straight Wall area (b3). Plate 9g shows larger scale photographs of the area round Bullialdus and Kies (e3), and of the Mare Humorum clefts and ridges (f3).

0317 G.M.T. 9.8.66. Moon's age 23·8 days. Diameter 25 inches. Note the Straight Wall (b3); it is about 105 kms. long and 800 feet high, and slopes down towards the East. The Riphaen Mts. (e6) are about 3,000 feet high.

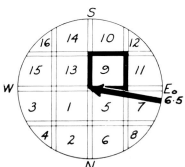

Plate 9b

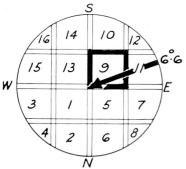

1758 G.M.T. 21.1.67. Moon's age 11·0 days. Diameter 25 inches. Note the clefts NE from Campanus (f2) and the ridges running down the right-hand side of the photograph. These are only visible when the lighting is very oblique. (See Plate 9e also.)

Plate 9c

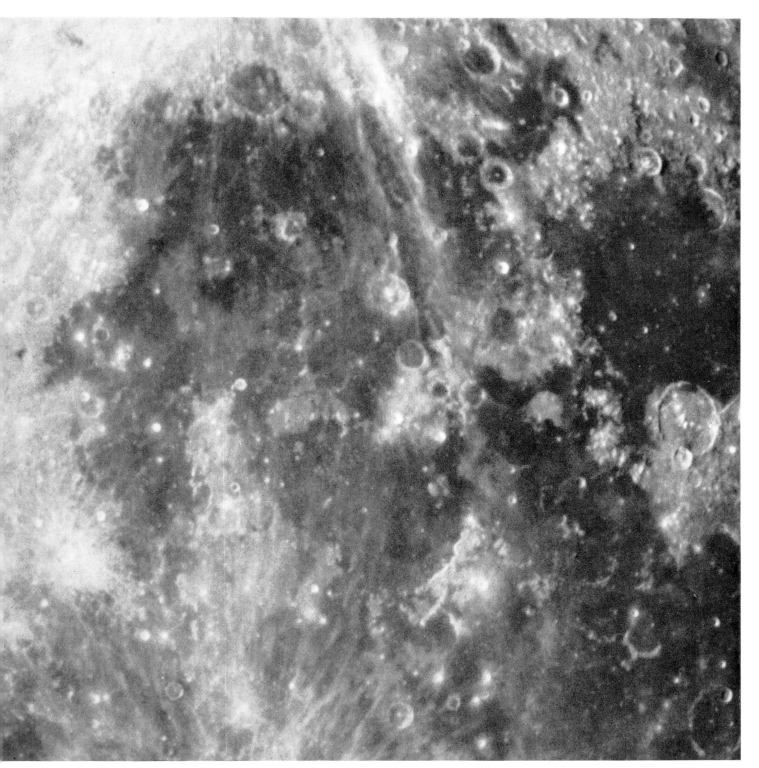

2135 G.M.T. 25.12.66. Moon's age 13·8 days. Diameter 25 inches. Compare this with Plates 9b and 9a, which show almost the same area under morning and afternoon lighting.

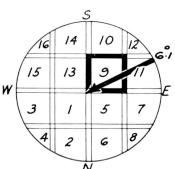

Plate 9d

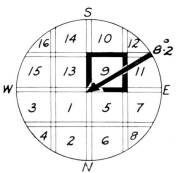

0516 G.M.T. 6.10.66. Moon's age 21·4 days. Diameter 25 inches. This photograph extends further to the Westward than the others in this group and overlaps into sections 1 and 13.

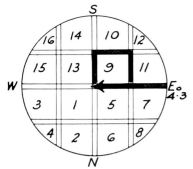

2236 G.M.T. 23.12.66. Moon's age 11·8 days. Diameter 25 inches. The Moon here is less than one day older than it is in Plate 9b and yet the clefts and ridges shown on the latter have virtually disappeared.

Plate 9f

0317 G.M.T. 9.8.66. Moon's age 23·8 days.

1928 G.M.T. 19.3.67. Moon's age 8·6 days.

Early morning (left) and late afternoon (right) views of the Straight Wall area. See Map **9** b3. The Moon's diameter in both these photographs is 36 inches approx.

Plate 9g

Clefts and ridges on the Western shores of the Mare Humorum. 1932 G.M.T. 21.3.67. Moon's age 10·6 days. Diameter 36 inches approx. Campanus and Mercator (top left) lie in Map **9** f2.

Bullialdus, Kies and the Kies dome, and the Hesiodus Cleft. 2056 G.M.T. 20.3.67. Moon's age 9·7 days. Diameter 36 inches approx. The Kies dome lies just under one diameter East from Kies, which is shown in Map **9** e2.

Map 10

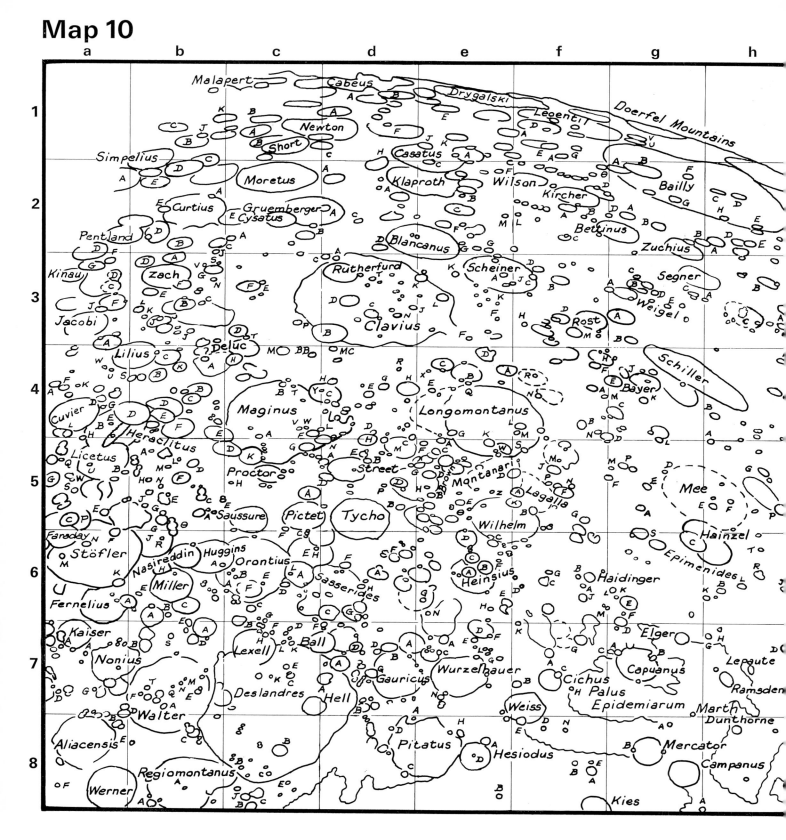

Crater Diameters

Mercator	38 kms.	(g8)
Werner	66 kms.	(a8)
Tycho	84 kms.	(d5)
Bettinus	66 kms.	(g2)
Moretus	105 kms.	(c2)

This map has been prepared from Plate 10a. Plates 10b, 10c and 10d sho
the same area under different lighting. Plate 10e shows larger sca
photographs of the area between Clavius (d3) and Bailly (g2).

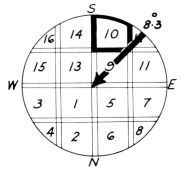

0609 G.M.T. 4.11.66. Moon's age 21·0 days. Diameter 25 inches. This is a favourable libration for this area.

Plate 10b

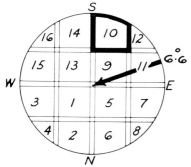

1758 G.M.T. 21.1.67. Moon's age 11·0 days. Diameter 25 inches. Compare this with Plate 10a; the libration here is not so favourable.

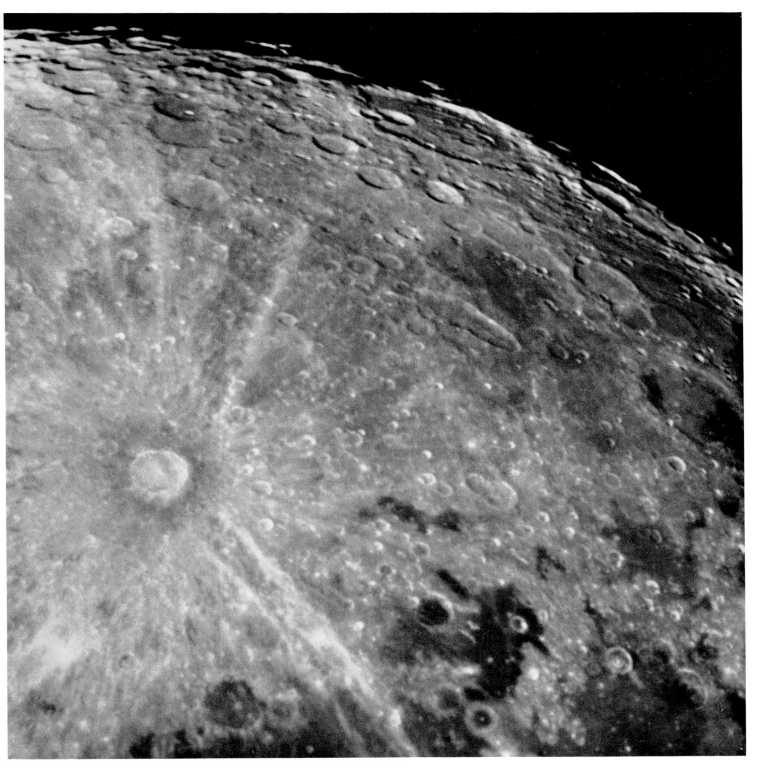

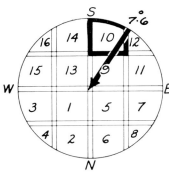

2104 G.M.T. 23.2.67. Moon's age 14·5 days. Diameter 25 inches. This was taken about 20 hours before Full Moon. Compare it with Plate 10d, which shows almost the same area.

Plate 10d

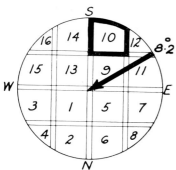

0516 G.M.T. 6.10.66. Moon's age 21·4 days. Diameter 25 inches. Compare this with Plates 10a and 10c.

Bailly. 2259 G.M.T. 5.3.66. Moon's age 13·5 days. Diameter 36 inches approx. Bailly lies in Map **10** g2.

Clavius to Bailly. 0516 G.M.T. 6.10.66. Moon's age 21·4 days. Diameter 36 inches approx. This is an enlargement of the part of Plate 10d. Compare the Bailly area here with the same region in the top photograph.

Map 11

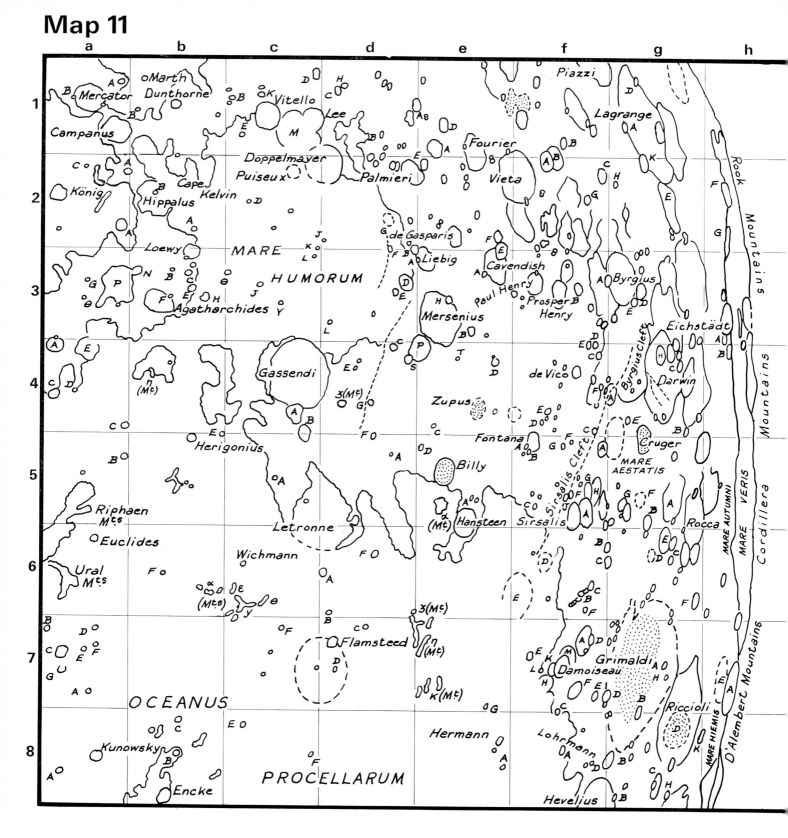

Crater Diameters

Damoiseau	35 kms.	(f7)
Kunowsky	21 kms.	(a8)
Billy	42 kms.	(e5)
Vieta	52 kms.	(f2)
Dunthorne	17½ kms.	(b1)

This map has been prepared from Plates 11a and 11c. Plates 11b and 11[
show the same area under different lighting. Plate 11e shows larger sca[
photographs of Darwin, the Sirsalis Cleft and Grimaldi (g4 to g7) and of th[
Mare Humorum and Gassendi (c3).

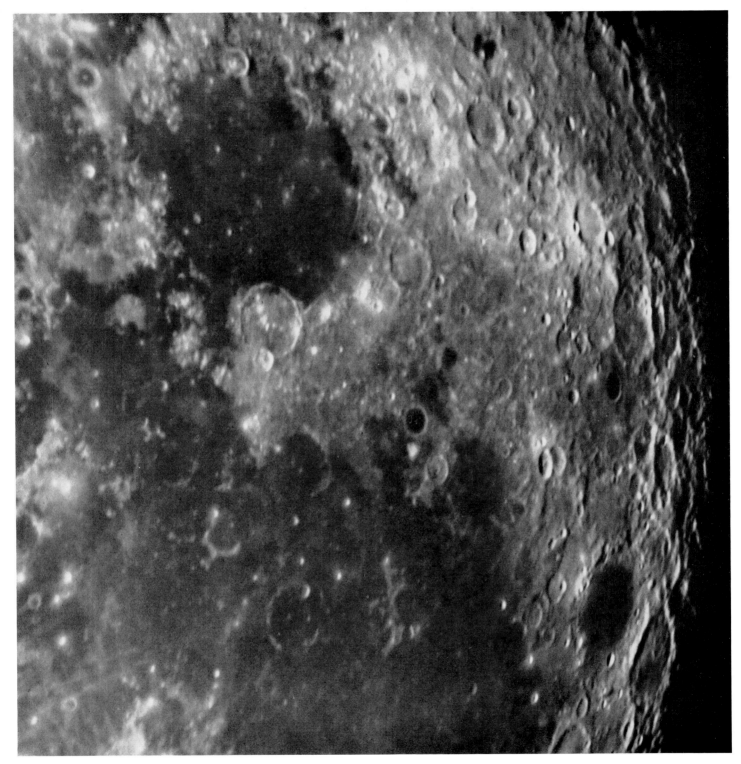

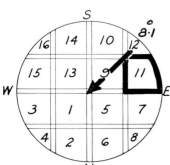

2310 G.M.T. 24.1.67. Moon's age 14·2 days. Diameter 25 inches. This is a fairly favourable libration for this area. Byrgius A (f3) may be seen as a "ray centre" on Plates 11b and 11c.

Plate 11b

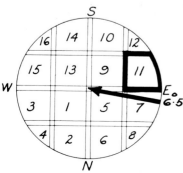

0317 G.M.T. 9.8.66. Moon's age 23·8 days. Diameter 25 inches. Compare this with Plate 11a. The "ray centre" (top right) is Byrgius A (f3).

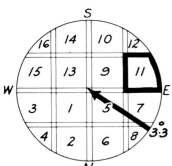

2229 G.M.T. 28.10.66. Moon's age 14·7 days. Diameter 25 inches. This was taken about 12 hours before Full Moon. Compare the detail on the limb here with that on Plate 11b. Although the libration in 11b is better than it is here, there is virtually nothing to be seen on the limb, since the lighting is not suitable.

Plate 11d

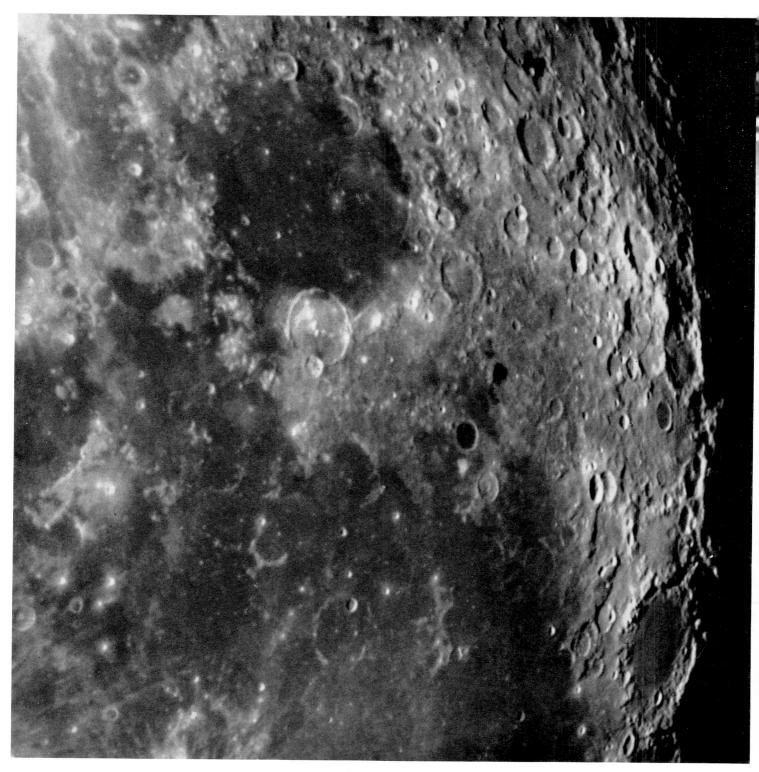

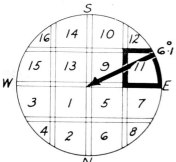

2036 G.M.T. 25.12.66. Moon's age 13·8 days. Diameter 25 inches. The Sirsalis Cleft (f5) will probably show up more clearly here if the reader turns the page 90° in a clockwise direction.

The Mare Humorum and Gassendi (c3). 2106 G.M.T. 20.4.67. Moon's age 11·0 days. Diameter 37 inches approx.

Darwin, the Sirsalis Cleft and Grimaldi. 2036 G.M.T. 25.12.66. Moon's age 13·8 days. Diameter 36 inches approx. This is an enlargement of part of Plate 11d. Note the "St. Andrew's Cross" marking on the NE wall of Grimaldi (g7).

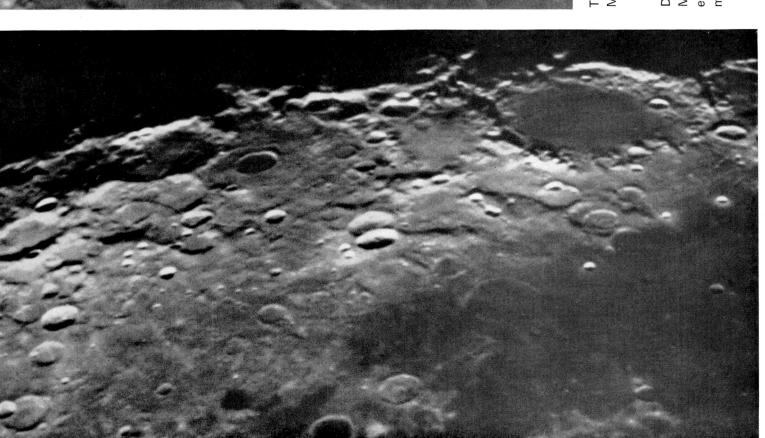

Map 12

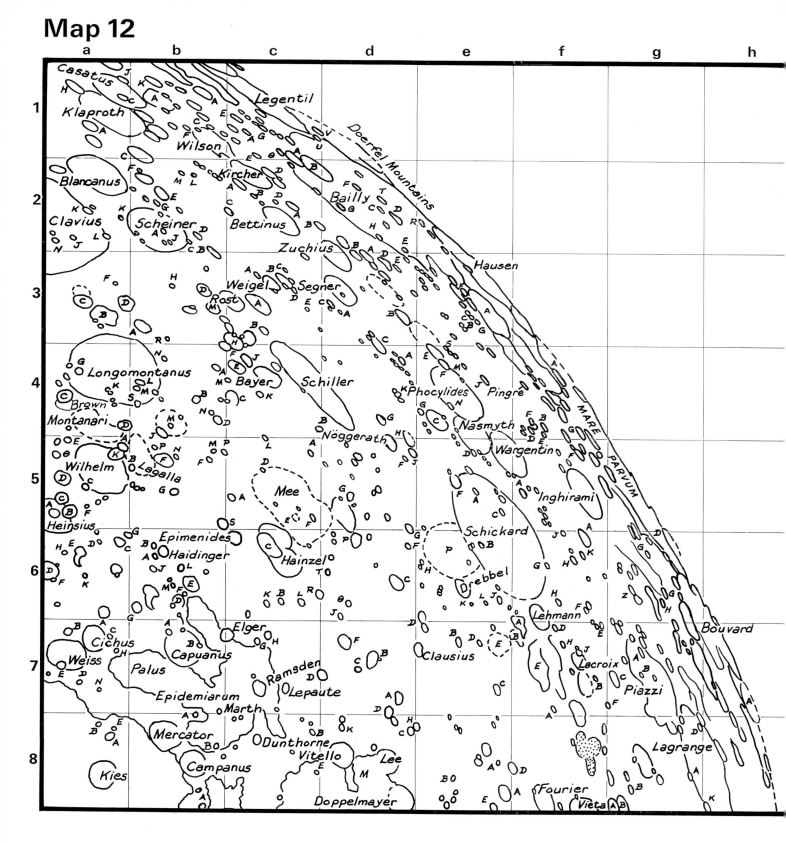

Crater Diameters

Lacroix	*38 kms. (f7)*
Vitello	*38 kms. (d8)*
Schickard	*202 kms. (e6)*
Zuchius	*64 kms. (d3)*
Casatus	*104 kms. (a1)*

This map has been prepared from Plate 12a. Plates 12b, 12c and 12d sho
the same area under different lighting. Plate 12e shows larger sca
photographs of Bailly (d2) and the area around Wargentin (e5).

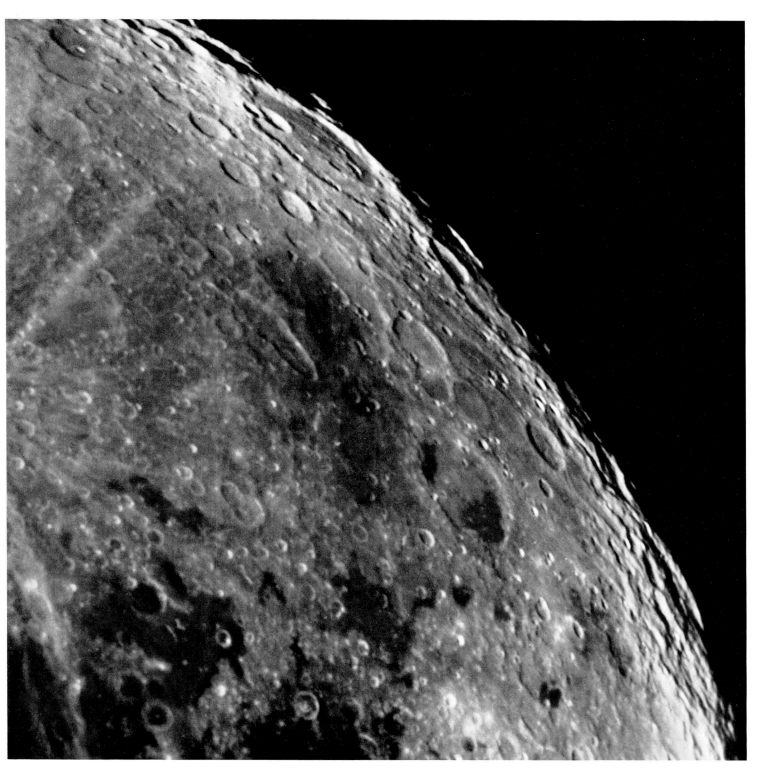

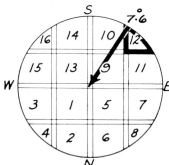

2104 G.M.T. 23.2.67. Moon's age 14·5 days. Diameter 25 inches. This is a good libration for this area. Bailly (d2) will not often be seen as well as this.

Plate 12b

2036 G.M.T. 25.12.66. Moon's age 13·8 days. Diameter 25 inches. Note that Wargentin (e5) has been filled in to form a plateau crater.

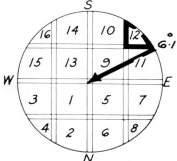

0317 G.M.T. 9.8.66. Moon's age 23·8 days. Diameter 25 inches. Compare this with Plate 12a, particularly near Bailly (d2).

Top right. Kircher (c2) to Schiller (c4). 1958 G.M.T. 21.4.67. Moon's age 11·9 days. Diameter 37 inches approx.

Plate 12d

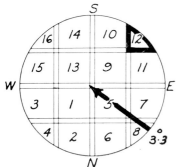

2229 G.M.T. 28.10.66. Moon's age 14·7 days. Diameter 25 inches. This was taken about 12 hours before Full Moon. Compare it with Plates 12a and 12c.

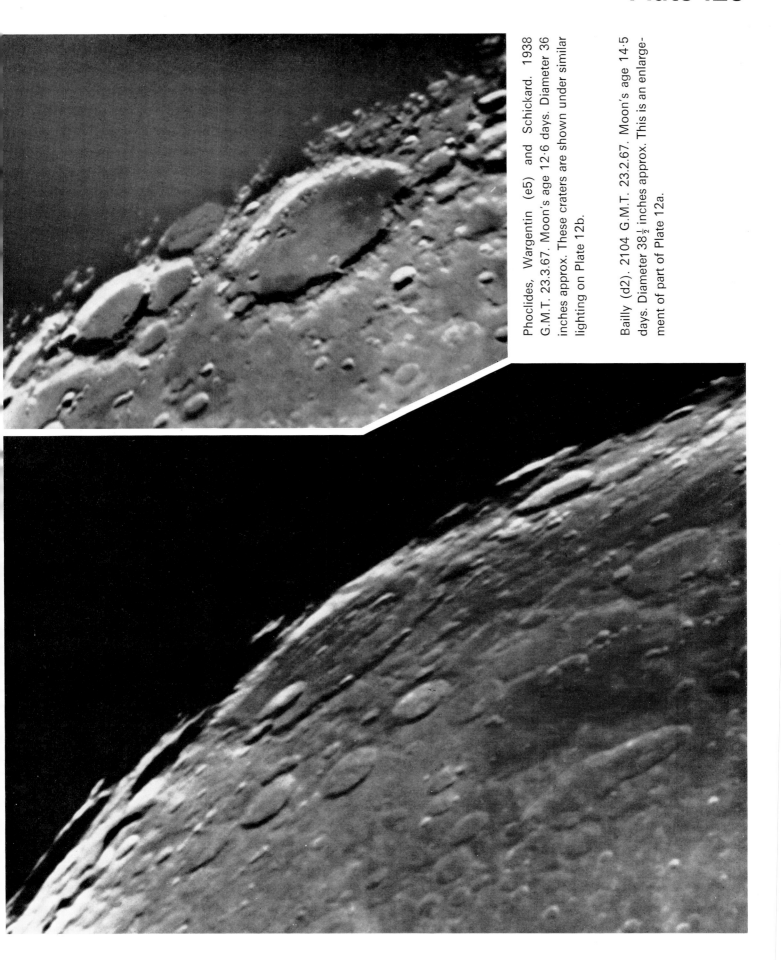

Phoclides, Wargentin (e5) and Schickard. 1938 G.M.T. 23.3.67. Moon's age 12·6 days. Diameter 36 inches approx. These craters are shown under similar lighting on Plate 12b.

Bailly (d2). 2104 G.M.T. 23.2.67. Moon's age 14·5 days. Diameter 38½ inches approx. This is an enlargement of part of Plate 12a.

Map 13

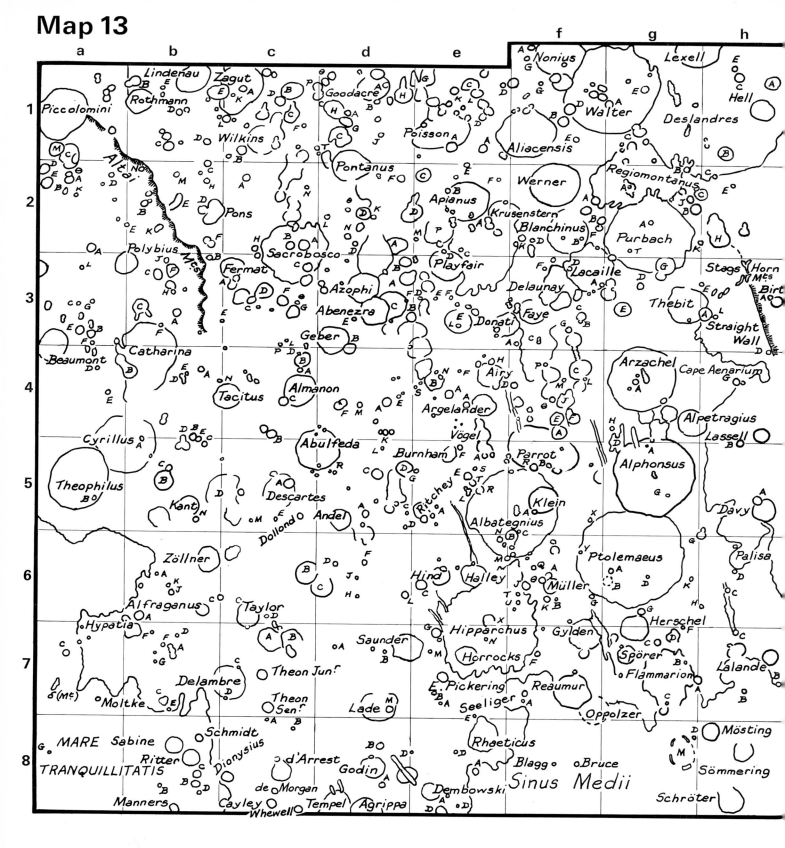

Crater Diameters

Mösting	26 kms. (h8)
Sabine	31 kms. (b8)
Argelander	42 kms. (e4)
Werner	66 kms. (f2)
Rothmann	42 kms. (b1)

This map has been prepared from Plate 13a. Plates 13b, 13c and 13d show the same area under different lighting. Plate 13e shows larger scale photographs of Abenezra (d3) and the area round Walter (g1). Plate 13f shows the area from Ptolemaeus (g6) to Hell (h1) on a larger scale. Plate 13g shows the Ptolemaeus–Alphonsus–Arzachel chain on a larger scale.

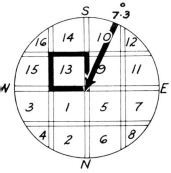

2046 G.M.T. 28.4.66. Moon's age 7·9 days. Diameter 25 inches. Note the small crater Regiomontanus A (g2) which lies on the summit of a mountain.

Plate 13b

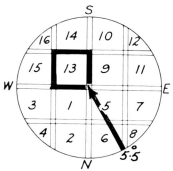

0215 G.M.T. 6.8.66. Moon's age 18·9 days. Diameter 25 inches. The Altai Mts. (b2) form an escarpment which is about 6,000 feet high generally; individual peaks may rise as much again.

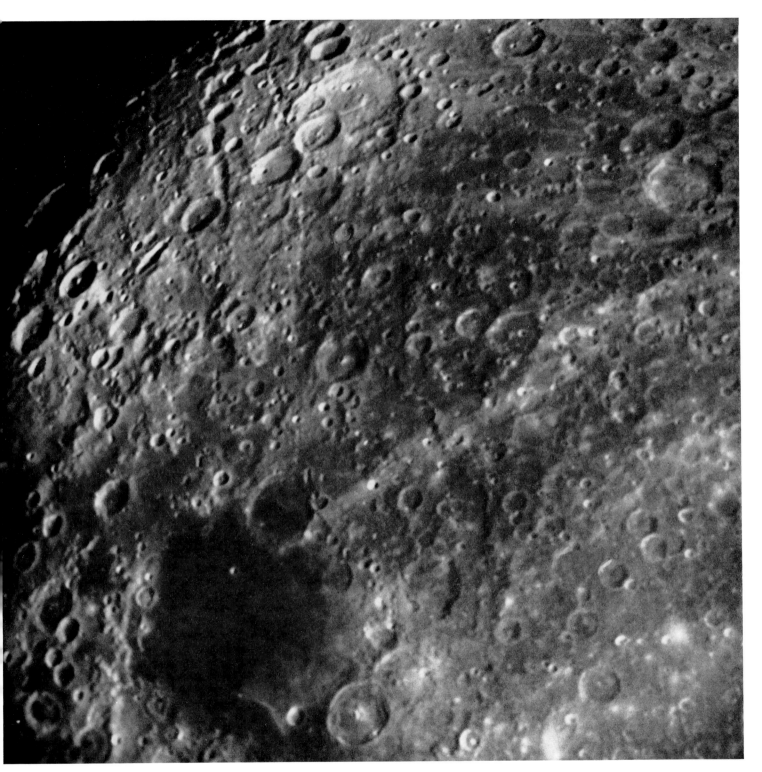

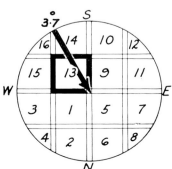

2334 G.M.T. 26.2.67. Moon's age 17·6 days. Diameter 25 inches. This photograph extends further to the South and West than the other members of this group. At the top it overlaps into the areas covered by Maps 14 and 16.

Plate 13d

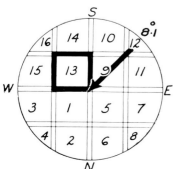

2314 G.M.T. 24.1.67. Moon's age 14·2 days. Diameter 25 inches. This was taken just over one day before Full Moon. Compare it with Plates 13a and 13b which show almost exactly the same area. Almost all the bright rays radiate from Tycho (Map **10** d5).

The area round Walter (g1). 1950 G.M.T. 17.4.67. Moon's age 7·9 days. Diameter 37 inches approx. Note the mountain-top crater Regiomontanus A (g2) and the dark bands in Stöfler (top left).

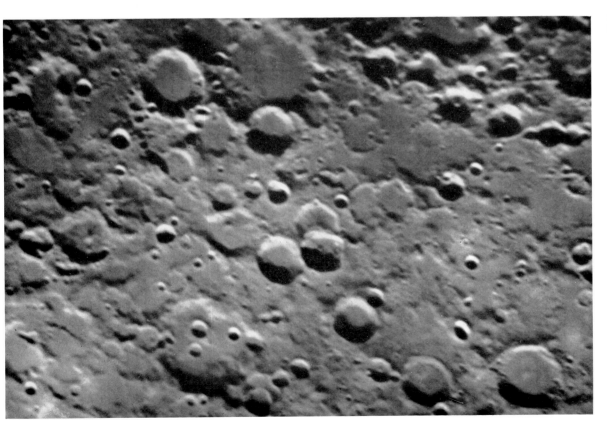

The area round Abenezra (d3). 2008 G.M.T. 16.5.67. Moon's age 7·2 days. Diameter 37 inches approx. Note the radial dark bands in Abenezra C.

Plate 13f

Ptolemaeus (g6) to Hell (h1). 1959 G.M.T. 19.3.67. Moon's age 8·6 days. Diameter 36 inches.

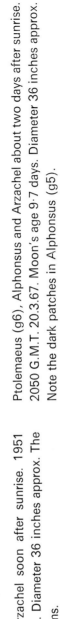

Ptolemaeus (g6), Alphonsus and Arzachel about two days after sunrise. 2050 G.M.T. 20.3.67. Moon's age 9·7 days. Diameter 36 inches approx. Note the dark patches in Alphonsus (g5).

Ptolemaeus (g6), Alphonsus and Arzachel soon after sunrise. 1951 G.M.T. 17.4.67. Moon's age 7·9 days. Diameter 36 inches approx. The diameter of Albategnius C (f5) is 7 kms.

Map 14

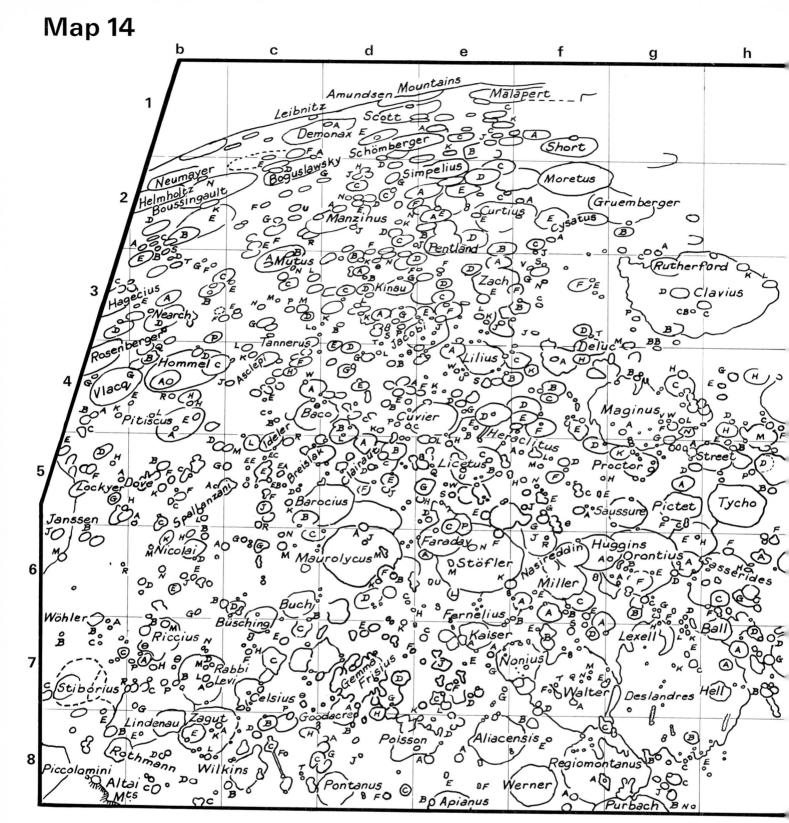

Crater Diameter

Walter	129 kms. (f7)
Stiborius	43½ kms. (a7)
Cuvier	66 kms. (e4)
Clavius	209 kms. (g3)
Manzinus	90 kms. (d2)

This map has been prepared from Plate 14a. Plates 14b, 14c and 14d sho▨ the same area under different lighting. Plate 14e shows larger sca▨ photographs of the area round Maurolycus (d6) and Stöfler (e6), an▨ Clavius (g3).

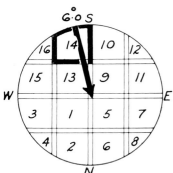

2122 G.M.T. 28.5.66. Moon's age 8·4 days. Diameter 25 inches. This is quite a good libration for the area at the top of the photograph.

Plate 14b

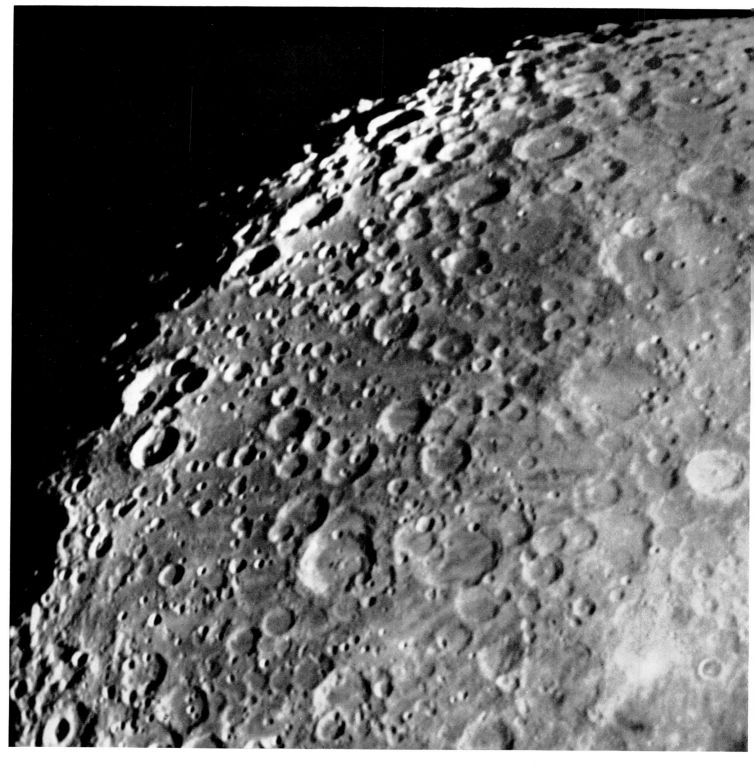

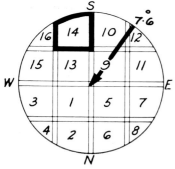

2336 G.M.T. 1.12.66. Moon's age 19·4 days. Diameter 25 inches. Compare this with Plates 14a and 14c, which show almost exactly the same area.

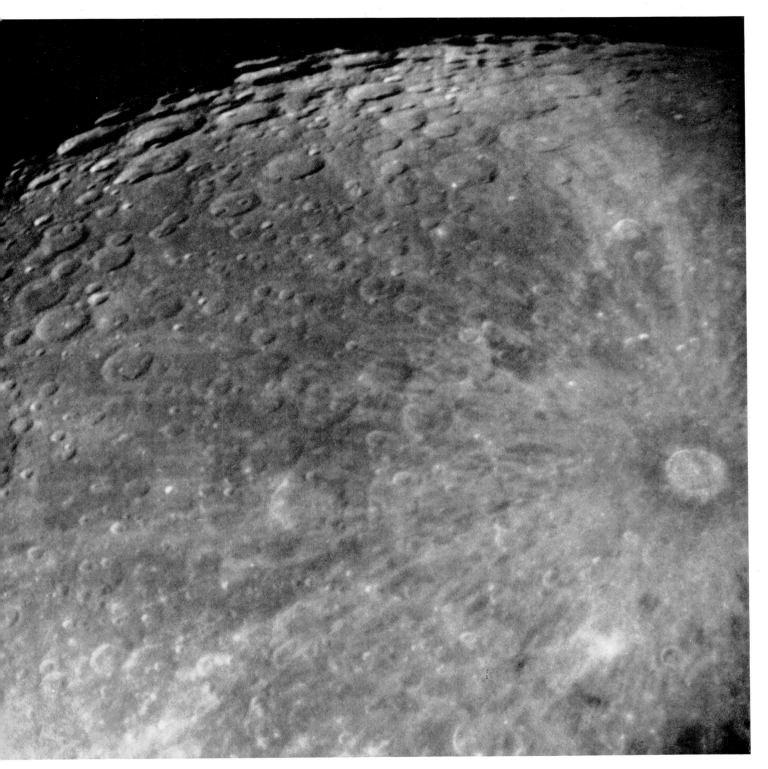

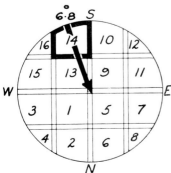

2326 G.M.T. 6.2.66. Moon's age 16·3 days. Diameter 25 inches. This was taken just over one day after Full moon. Compare it with Plates 14a and 14b, which show almost exactly the same area.

Plate 14d

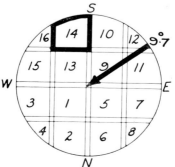

1959 G.M.T. 19.3.67. Moon's age 8·6 days. Diameter 25 inches. Compare this with Plate 14a. The Moon's age is very nearly the same in both cases but the librations are quite different.

The area round Clavius (g3). 2046 G.M.T. 20.3.67. Moon's age 9·7 days. Moon's diameter 36 inches approx. The small crater Clavius CB (g3) is 7 kms. in diameter.

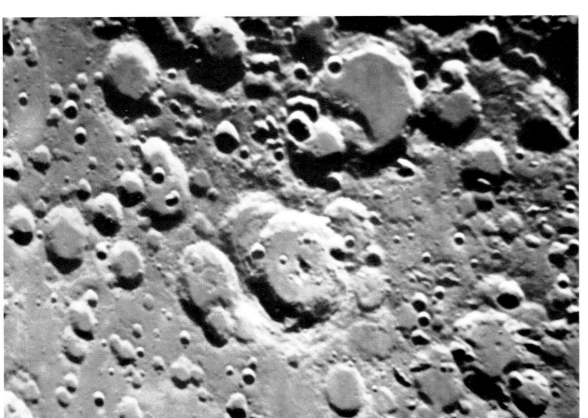

The area round Maurolycus (d6) and Stöfler (e6). 2021 G.M.T. 16.5.67. Moon's age 7·3 days. Diameter 37 inches approx. Note that both Maurolycus (d6) and Ideler (c5) are overlapping smaller craters.

Map 15

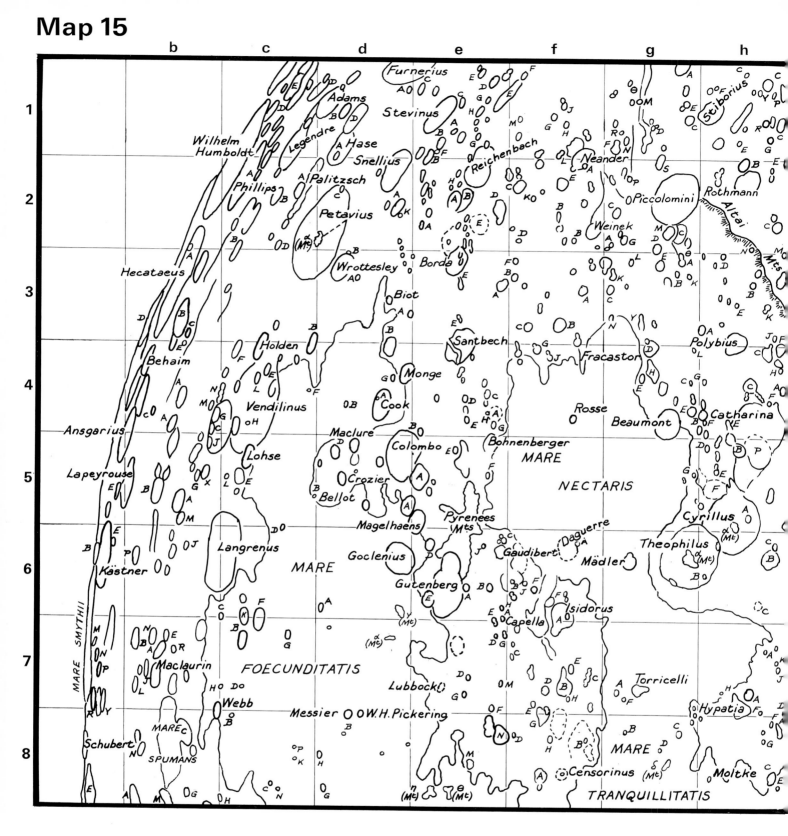

Labels on map (selected):

Furnerius, Adams, Stevinus, Reichenbach, Neander, Piccolomini, Rothmann, Altai Mts, Wilhelm Humboldt, Legendre, Hase, Snellius, Phillips, Palitzsch, Weinek, Petavius, Wrottesley, Borda, Hecataeus, Biot, Santbech, Fracastor, Polybius, Behaim, Holden, Monge, Cook, Rosse, Beaumont, Catharina, Vendilinus, Maclure, Colombo, Bohnenberger, MARE NECTARIS, Lohse, Crozier, Bellot, Ansgarius, Magelhaens, Pyrenees (Mts), Cyrillus, Lapeyrouse, Goclenius, Gaudibert, Daguerre, Theophilus, Mädler, Langrenus, Gutenberg, Isidorus, Kästner, MARE, Capella, Torricelli, Maclaurin, FOECUNDITATIS, Lubbock, Hypatia, Webb, Messier, W.H. Pickering, Schubert, MARE SPUMANS, MARE SMYTHII, Censorinus (Mt), MARE TRANQUILLITATIS, Moltke

Crater Diameters		
Moltke	7 kms.	(h8)
Webb	26 kms.	(b8)
Theophilus	101 kms.	(g6)
Piccolomini	80 kms.	(g2)
Stevinus	70 kms.	(e1)

This map has been prepared from Plates 15a and 15d. Plates 15b, 15c and 15e show the same area under different lighting. Plate 15e also shows a larger scale photograph of the area round William Humboldt (c2).

2152 G.M.T. 27.4.66. Moon's age 6·9 days. Diameter 25 inches. Compare this with the other members of this group, which show the detail near the limb much more clearly.

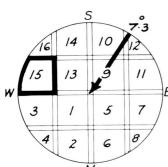

Plate 15b

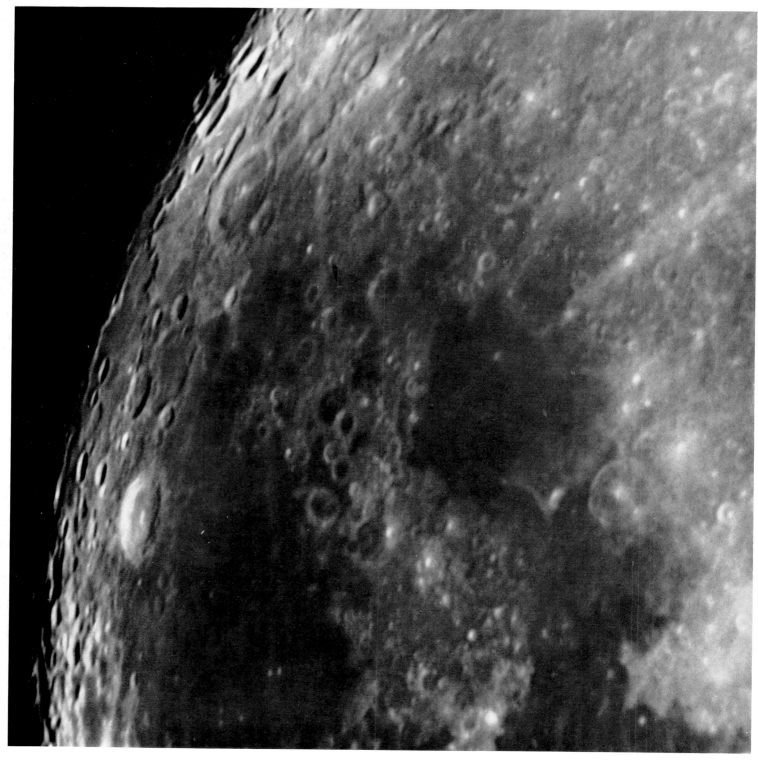

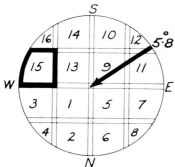

2125 G.M.T. 28.11.66. Moon's age 16·3 days. Diameter 25 inches. This was taken about 19 hours after Full Moon.

2228 G.M.T. 8.1.66. Moon's age 17·0 days. Diameter 25 inches. Compare this with Plate 15b which shows almost exactly the same area.

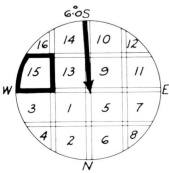

Plate 15d

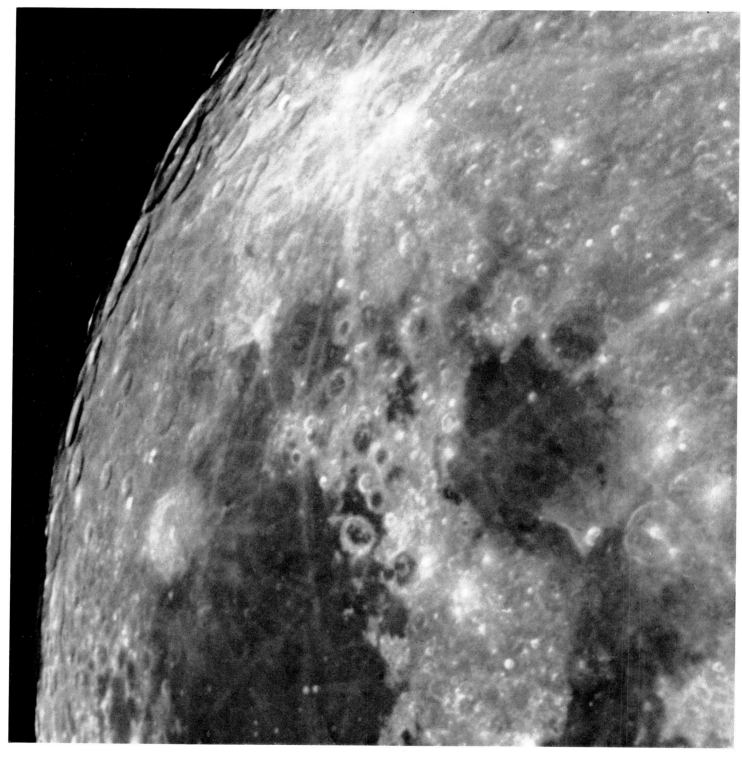

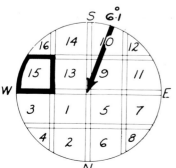

2300 G.M.T. 24.2.67. Moon's age 15·6 days. Diameter 25 inches. This was taken about 5 hours after Full Moon. The libration is not favourable, but even so the craters on the limb near William Humboldt (c2) will not often be seen like this.

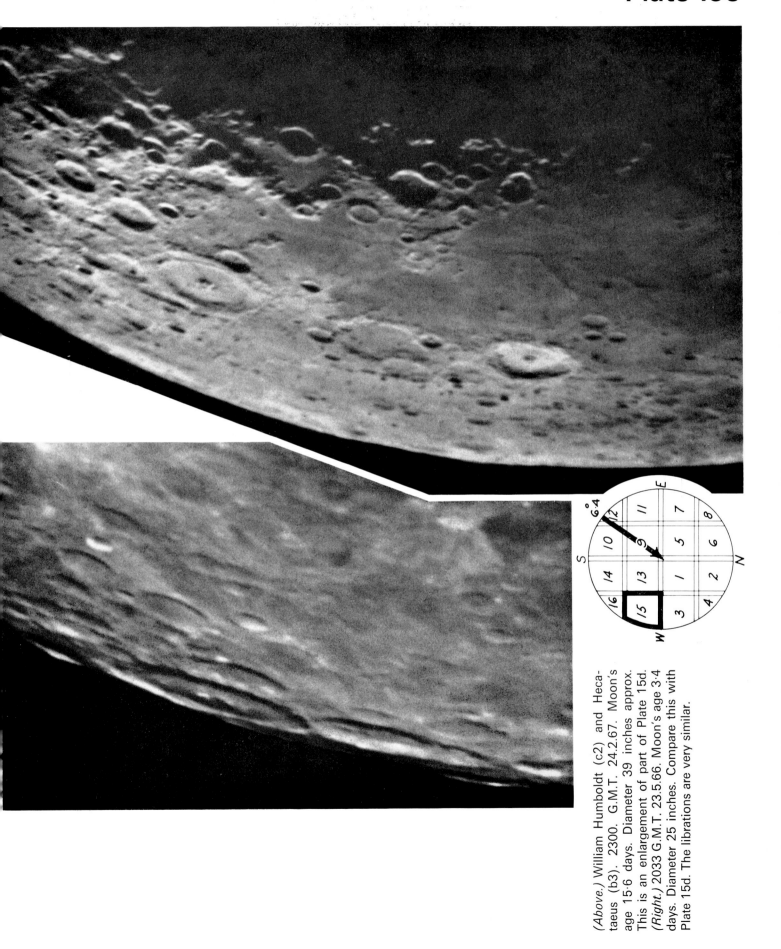

(*Above.*) William Humboldt (c2) and Hecataeus (b3). 2300. G.M.T. 24.2.67. Moon's age 15·6 days. Diameter 39 inches approx. This is an enlargement of part of Plate 15d.
(*Right.*) 2033 G.M.T. 23.5.66. Moon's age 3·4 days. Diameter 25 inches. Compare this with Plate 15d. The librations are very similar.

Map 16

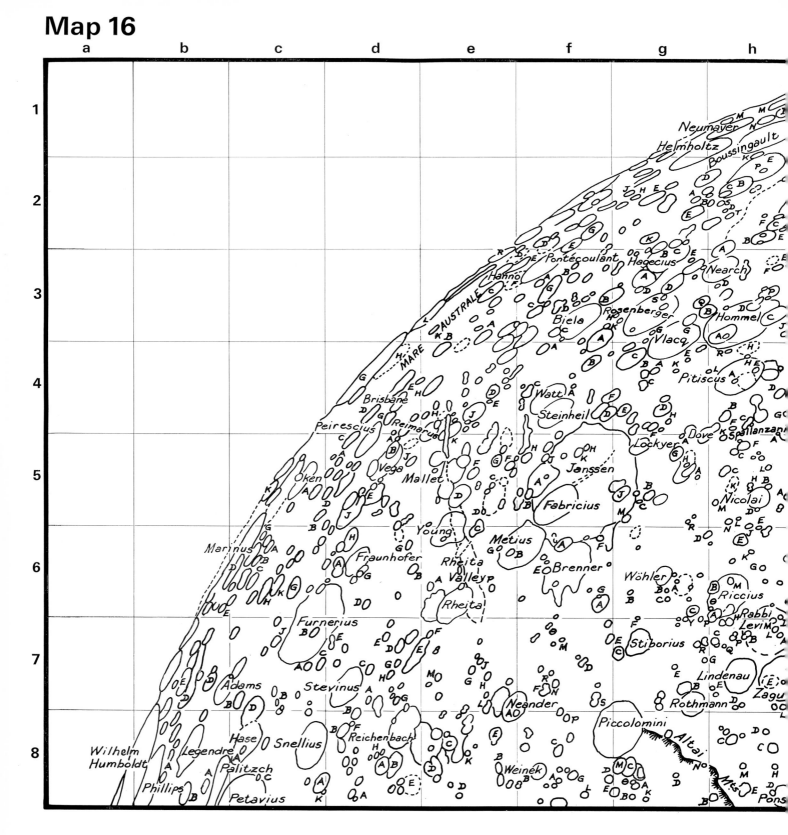

Crater Diameters

Weinek	30 kms. (f8)
Snellius	80 kms. (c8)
Steinheil	70 kms. (f4)
Boussingault	78 kms. (h1)
Brisbane	47 kms. (d4)

This map has been prepared from Plates 16a and 16c. Plates 16b, 16d ar
16e show the same area under different lighting. Plate 16f shows a larg
scale photograph of the Mare Australe area (d4).

2109 G.M.T. 23.6.66. Moon's age 5·0 days. Diameter 25 inches. Compare this with Plate 16c, which shows almost exactly the same area.

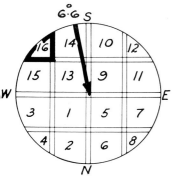

Plate 16b

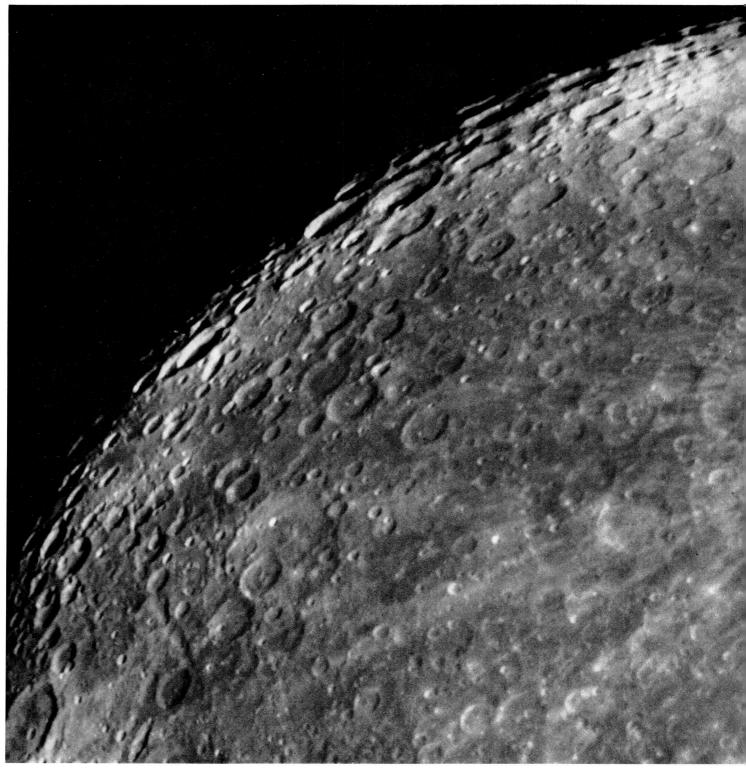

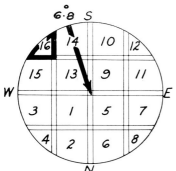

2326 G.M.T. 6.2.66. Moon's age 16·3 days. Diameter 25 inches. This was taken nearly 1½ days after Full Moon, and so some of the limb detail has already disappeared, despite the favourable libration.

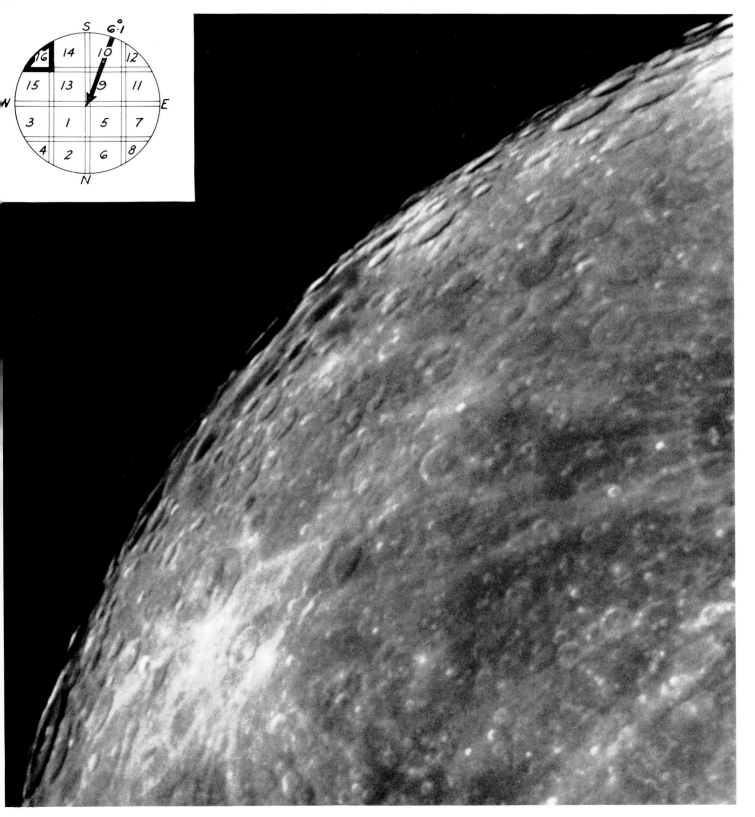

2259 G.M.T. 24.2.67. Moon's age 15·6 days. Diameter 25 inches. This was taken about 5 hours after Full Moon. The libration was not favourable, but even so the craters on the limb near Brisbane (d4) will not often be seen like this.

Plate 16d

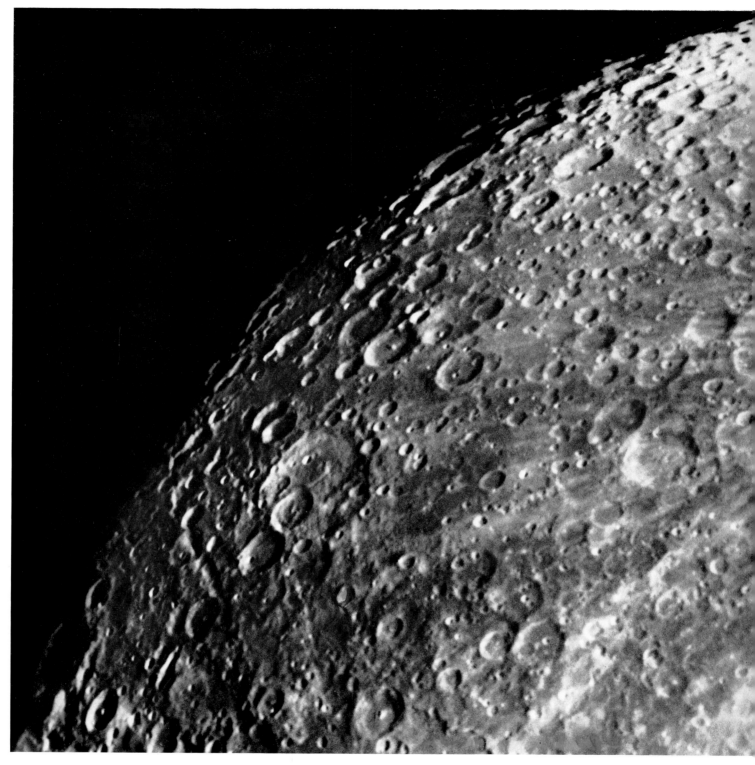

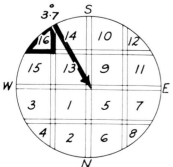

2334 G.M.T. 26.2.67. Moon's age 17·6 days. Diameter 25 inches. Compare this with Plate 16b, which shows almost exactly the same area.

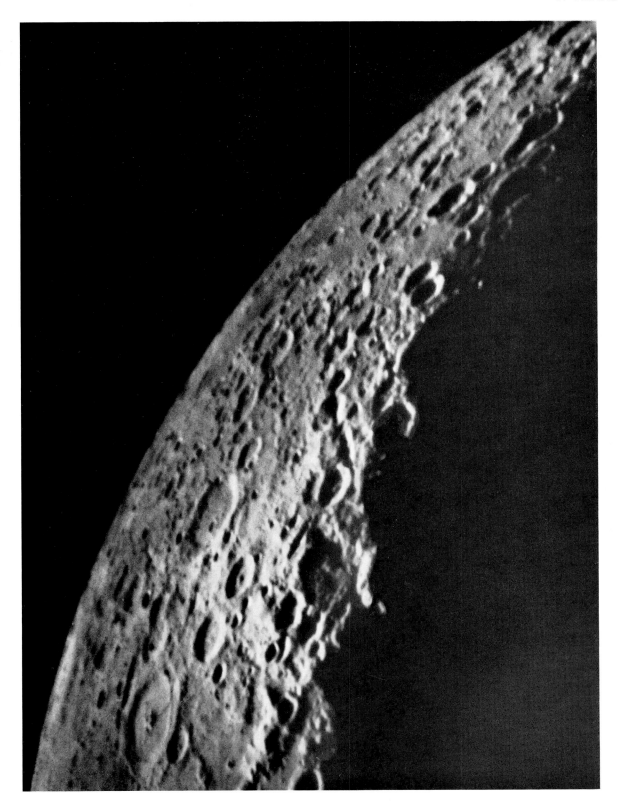

2036 G.M.T. 23.5.66. Moon's age 3·4 days. Diameter 25 inches. Compare this with Plate 16c. The librations are very similar.

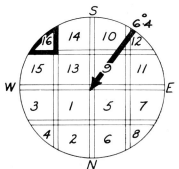

Plate 16f

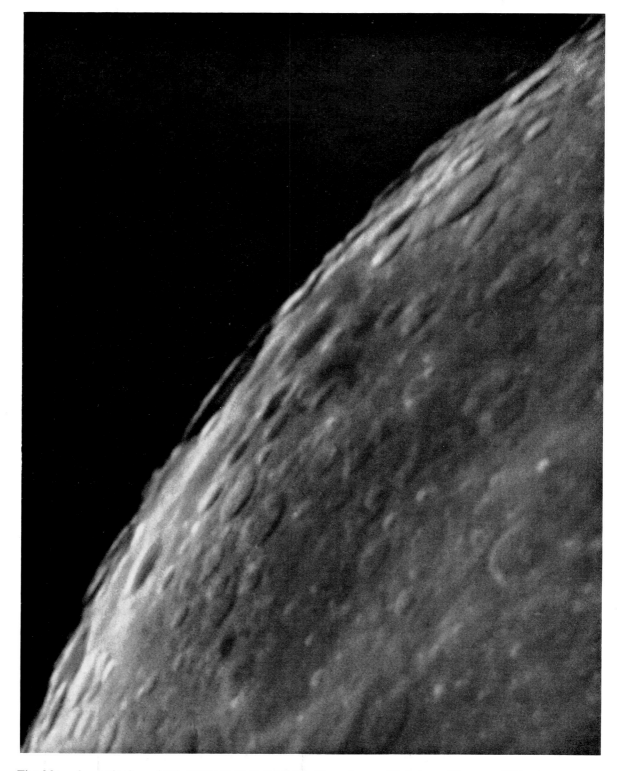

The Mare Australe Area (d4). 2259 G.M.T. 24.2.67. Moon's age 15·6 days. Diameter 39 inches approx. This is an enlargement of part of Plate 16c.

Plate 17

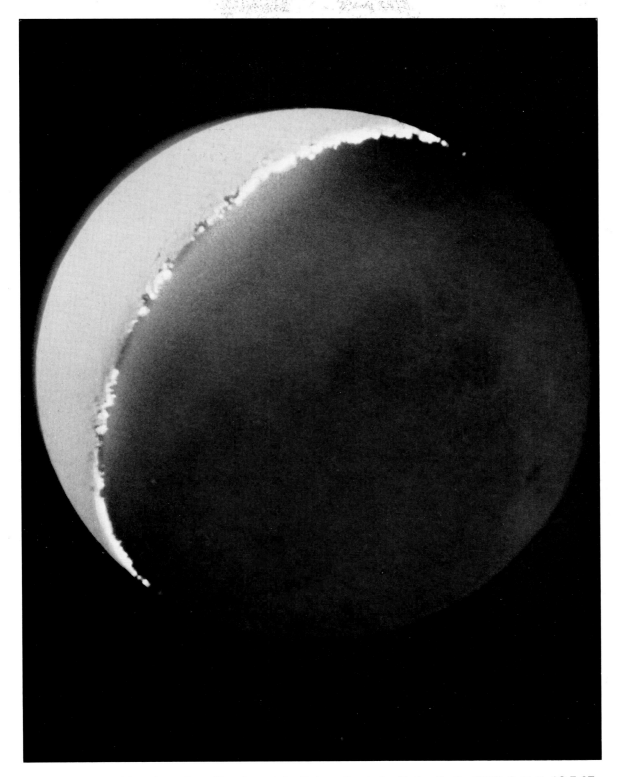

Earthshine—the Moon's surface illuminated by light reflected off the Earth. 2032 G.M.T. 13.5.67. Moon's age 4·3 days. Compare this with the Full Moon Key photograph, immediately preceding Map 1. The Earthshine needed an exposure of 10 seconds, and the part of the Moon illuminated by the Sun was therefore grossly over-exposed.

TABLE OF EXPOSURES

Date	G.M.T.	Focal Ratio	Emulsion	Exposure (seconds)	Moon's Age (days)	Plate Nos.
8-11-65*	2132	f29	Kodak O.250 Plate	0·4	15·3	7e
8- 1-66	2228	f24	Kodak O.250 Plate	0·3†	17·0	3b, 15c
31- 1-66	1947	f24	Kodak O.250 Plate	0·3	10·1	5b
2- 2-66	1944	f24	Kodak O.250 Plate	0·25	12·1	7b, 8b
6- 2-66	2326	f24	Kodak O.250 Plate	0·2	16·3	14c, 16b
5- 3-66	2259	f41	Ilford Zenith Plate	0·1	13·5	10e
27- 4-66	2152	f24	Ilford G.30 Plate	0·5	6·9	15a
28- 4-66	2046	f30	Ilford G.30 Plate	0·8	7·9	1a, 13a
23- 5-66	2033	f24	Kodak O.250 Plate	0·7	3·4	15e
23- 5-66	2034	f24	Kodak O.250 Plate	0·7	3·4	3e, 4b
23- 5-66	2036	f24	Kodak O.250 Plate	0·7	3·4	16e
28- 5-66	2122	f30	Ilford G.30 Plate	0·8	8·4	14a
29- 5-66	2103	f30	Ilford G.30 Plate	0·8	9·4	2c
23- 6-66	2109	f24	Ilford G.30 Plate	1·0	5·0	16a
6- 8-66	0211	f30	Ilford G.30 Plate	0·7	18·9	2b
6- 8-66	0215	f30	Ilford G.30 Plate	0·7	18·9	1b, 13b
9- 8-66	0315	f30	Ilford G.30 Plate	1·1	23·8	5a, 6a
9- 8-66	0317	f30	Ilford G.30 Plate	1·1	23·8	9a, 9f, 11b, 12c
6-10-66	0516	f30	Ilford G.30 Plate	1·5	21·4	9d, 10d, 10e
28-10-66	2219	f30	Ilford G.30 Plate	0·4	14·7	7e, 8c
28-10-66	2229	f30	Ilford G.30 Plate	0·4	14·7	11c, 12d
29-10-66	2248	f30	Ilford G.30 Plate	0·3	15·7	4d
4-11-66	0609	f30	Ilford G.30 Plate	1·5	21·0	10a
22-11-66	1814	f30	Ilford G.30 Plate	0·8	10·2	2a
23-11-66	2142	f30	Ilford G.30 Plate	0·7	11·3	6c
28-11-66	2125	f30	Ilford G.30 Plate	0·5	16·3	15b
28-11-66	2127	f30	Ilford G.30 Plate	0·5	16·3	4c
1-12-66	2336	f30	Ilford G.30 Plate	0·7	19·4	14b
7-12-66	0655	f30	Ilford G.30 Plate	2·5‡	24·6	7c, 8d
23-12-66	2232	f30	Ilford G.30 Plate	0·6	11·8	6d
23-12-66	2236	f30	Ilford G.30 Plate	0·6	11·8	9e
25-12-66	2036	f30	Ilford G.30 Plate	0·4	13·8	7d, 11d, 11e, 12b
25-12-66	2040	f30	Ilford G.30 Plate	0·4	13·8	6b, 8a
25-12-66	2135	f30	Ilford G.30 Plate	0·4	13·8	5c, 9c
25-12-66	2137	f30	Ilford G.30 Plate	0·4	13·8	1c
21- 1-67	1758	f30	Ilford G.30 Plate	0·5	11·0	9b, 10b
24- 1-67	2310	f30	Ilford G.30 Plate	0·4	14·2	7a, 11a
24- 1-67	2314	f30	Ilford G.30 Plate	0·4	14·2	13d
16- 2-67	1757	f30	Ilford G.30 Plate	1·0	7·3	3a, 4a
19- 2-67	1801	f30	Gevaert R.23 Plate	0·5	10·3	5d
23- 2-67	2104	f30	Ilford G.30 Plate	0·6†	14·5	10c, 12a, 12e
24- 2-67	2259	f30	Ilford G.30 Plate	0·6†	15·6	16c, 16f
24- 2-67	2300	f30	Ilford G.30 Plate	0·6†	15·6	15d, 15e
24- 2-67	2353	f30	Ilford G.30 Plate	0·6†	15·6	3d
26- 2-67	2334	f30	Ilford G.30 Plate	0·6†	17·6	13c, 16d
26- 2-67	2336	f30	Ilford G.30 Plate	0·6†	17·6	3c
18- 3-67	1835	f30	Ilford G.30 Plate	1·0	7·6	1d, 2d
19- 3-67	1925	f60	Ilford F.P.3 35 mm.	1	8·6	2e
19- 3-67	1928	f60	Ilford F.P.3 35 mm.	1	8·6	9f
19- 3-67	1959	f30	Ilford G.30 Plate	0·8	8·6	13f, 14d
20- 3-67	2029	f60	Ilford F.P.3 35 mm.	1	9·7	6e
20- 3-67	2041	f60	Ilford F.P.3 35 mm.	1	9·7	5e
20- 3-67	2046	f60	Ilford F.P.3 35 mm.	1	9·7	14e
20- 3-67	2050	f60	Ilford F.P.3 35 mm.	½	9·7	13g
20- 3-67	2056	f60	Ilford F.P.3 35 mm.	1	9·7	9g
21- 3-67	1908	f60	Ilford F.P.3 35 mm.	1	10·6	5e

Date	G.M.T.	Focal Ratio	Emulsion	Exposure (seconds)	Moon's Age (days)	Plate Nos.
21- 3-67	1923	f60	Ilford F.P.3 35 mm.	1	10·6	6e
21- 3-67	1932	f60	Ilford F.P.3 35 mm.	1	10·6	9g
23- 3-67	1938	f60	Ilford F.P.3 35 mm.	1	12·6	12e
23- 3-67	1944	f60	Ilford F.P.3 35 mm.	1	12·6	8e
17- 4-67	1950	f52	Ilford Pan.F. 35 mm.	2	7·9	13e
17- 4-67	1951	f52	Ilford Pan.F. 35 mm.	2	7·9	13g
20- 4-67	2106	f52	Ilford Pan.F. 35 mm.	2	11·0	11e
21- 4-67	1952	f52	Ilford Pan.F. 35 mm.	½	11·9	8e
21- 4-67	1958	f52	Ilford Pan.F. 35 mm.	2	11·9	12c
21- 4-67	2033	f52	Ilford Pan.F. 35 mm.	2	12·0	8e
23- 4-67	2150	f52	Ilford Pan.F. 35 mm.	⅛	14·0	8e
13- 5-67	2002	f52	Ilford Pan.F. 35 mm.	3	4·2	3g
13- 5-67	2032	f7·25§	Ilford Pan.F. 35 mm.	10	4·3	17
15- 5-67	2000	f52	Ilford Pan.F. 35 mm.	3	6·2	3f
15- 5-67	2008	f52	Ilford Pan.F. 35 mm.	2	6·2	3g
15- 5-67	2015	f52	Ilford Pan.F. 35 mm.	3	6·2	3f
16- 5-67	2008	f52	Ilford Pan.F. 35 mm.	2	7·2	13e
16- 5-67	2012	f52	Ilford Pan.F. 35 mm.	3	7·2	2e
16- 5-67	2016	f52	Ilford Pan.F. 35 mm.	3	7·2	1e
16- 5-67	2017	f52	Ilford Pan.F. 35 mm.	3	7·2	1e
16- 5-67	2021	f52	Ilford Pan.F. 35 mm.	2	7·3	14e
20- 5-67	2110	f52	Ilford Pan.F. 35 mm.	½	11·3	8e
20- 8-67	2304	f30	Ilford G.30 Plate	0·3	14·9	3e, 4e

Except where noted the instrument was a 12-inch Newtonian Reflector with a silvered mirror.
* The instrument here was a 6-inch Newtonian Reflector with a silvered mirror.
† An Ilford α pale yellow filter was used here.
‡ This was taken through mist; hence the long exposure.
§ This was taken at the Prime Focus.

Emulsion speeds:	Ilford F.P.3	A.S.A. 120
	Ilford Pan.F.	A.S.A. 50
	Ilford Zenith	A.S.A. 50
	Ilford G.30	A.S.A. 10
	Kodak O.250	A.S.A. 25
	Gevaert R.23	A.S.A. 12

These are the maker's quoted speeds. They do not really apply to the fairly long exposures used here.

INDEX OF NAMED FORMATIONS

This index contains all the formations listed in *Named Lunar Formations* by Mary A. Blagg and K. Müller (1935). A few additions are indicated with an asterisk.(*)

Formations such as Mares, Capes and Mountains are indexed under their proper names. The groups of figures and letters against the formations indicate where they will be found on the various key maps. Thus the group **3 b4** against Alhazen indicates that it lies in square b4 on Map **3**; Adams, **15 d1, 16 b7**, may be found on Maps **15** and **16**.

Cook	**15** d4
Copernicus	**5** e4
Cordillera Mts.	**11** h5
Crisium, Mare	**3** c5, **4** c1
Crozier	**15** d5
Crüger	**11** g5
Curtius	**10** b2, **14** e2
Cusanus	**2** c8
Cuvier	**10** a4, **14** e4
Cyrillus	**13** a5, **15** h5
Cysatus	**10** c2, **14** f2

D

Daguerre	**15** f6
D'Alembert Mts.	**7** h2, **11** h7
Damoiseau	**7** f2, **11** f7
Daniell	**1** b8, **2** a2, **4** g4
Darney	**9** e5
D'Arrest	**1** d1, **13** c8
Darwin	**11** g4
Da Vinci	**3** d3
Davy	**9** b5, **13** h5
Dawes	**1** b4, **3** g5
Debes	**3** d7, **4** d3
Dechen	**8** d5
De Gasparis	**11** e2
Delambre	**13** c7
De la Rue	**2** a6, **4** h7
Delaunay	**13** f3
Delisle	**5** g8, **6** g2, **8** b3
Delmotte	**3** c7, **4** c2
Deluc	**10** c4, **14** f3
Dembowski	**1** f2, **13** e8
Democritus	**2** c6
Demonax	**14** c1
De Morgan	**1** d1, **13** c8
Descartes	**13** c5
Deseilligny	**1** c5, **3** h6, **4** h1
Deslandres*	**9** b1, **10** c7, **13** h1, **14** g7
De Vico	**11** f4
Deville, Cape	**2** f4
Dionysius	**1** d1, **13** c8
Diophantus	**5** g8, **6** g1, **7** a8, **8** b2
Doerfel Mts.	**10** g1, **12** d1
Dollond	**13** c5
Donati	**13** e3
Doppelmayer	**9** h3, **11** c2, **12** d8
Dove	**14** b5, **16** g5
Draper	**5** e5
Drebbel	**12** e6
Drygalski*	**10** e1
Dunthorne	**9** g2, **10** h8, **11** b1, **12** c8

E

Egede	**2** e4
Eichstädt	**11** h3
Eimmart	**3** b6, **4** c2
Elger	**9** f1, **10** g7, **12** c7
Encke	**7** c4, **11** b8
Endymion	**4** g7
Epidemiarum, Palus	**9** f2, **10** g7, **12** b7
Epigenes	**2** g7, **6** b7
Epimenides	**10** g6, **12** b6
Eratosthenes	**5** c5
Euclides	**7** a1, **11** a6
Euctemon	**2** e8
Eudoxus	**2** d4
Euler	**5** g7, **8** a2

F

Fabricus	**16** f5
Faraday	**10** a6, **14** e6
Fauth	**5** e3
Faye	**13** f3
Fermat	**13** c3
Fernelius	**10** a6, **14** e6
Feuillé	**2** h1, **5** c7, **6** b1
Firmicus	**3** b3
Flammarion	**5** b1, **13** g7
Flamsteed	**7** d2, **9** h8, **11** d7
Foecunditatis, Mare	**3** c1, **15** c7
Fontana	**11** f5
Fontenelle	**2** h6, **6** c7
Foucault	**6** f5, **8** a6
Fourier	**11** e1, **12** f8
Frascator	**15** g4
Fra Mauro	**5** e1, **9** c6
Franklin	**4** e4
Franz	**3** e5
Fraunhöfer	**16** d6
Fresnel, Cape	**1** f7, **2** e1
Frigoris, Mare	**2** e5, **6** c6
Furnerius	**15** d1, **16** c7

G

Galilaei	**7** f5
Galle	**2** d6
Galvani	**8** e6
Gambart	**5** d2, **9** c8
Gärtner	**2** c6
Gassendi	**9** h5, **11** c4
Gaudibert	**15** e6
Gauricus	**9** c1, **10** d7
Gauss	**3** c8, **4** c4
Gay-Lussac	**5** e5
Gay-Lussac, Sinus	**5** e5
Geber	**13** d3
Geminus	**3** d8, **4** d4
Gemma Frisius	**14** d7
Gerard	**8** e5
Gioja	**2** f8
Glaisher	**3** c4
Goclenius	**15** d6
Godin	**1** e1, **13** d8
Goldschmidt	**2** f7, **6** b8
Goodacre	**13** d1, **14** d8
Gould	**9** d4
Grimaldi	**7** g1, **11** g7
Groves	**1** b8, **2** a3, **4** g5
Gruemberger	**10** c2, **14** g2
Gruithuisen	**6** g2, **8** c3
Guericke	**9** c5
Gutenberg	**15** e6
Gylden	**13** f7

H

Hadley, Mt.	**1** f6
Haemus Mts.	**1** c4, **3** h5
Hagecius	**14** a3, **16** g3
Hahn	**3** c7, **4** c3
Haidinger	**10** f6, **12** b6
Hainzel	**10** h6, **12** c6
Hall	**1** a7, **3** f8, **4** f3
Halley	**13** e6
Hanno	**16** e3
Hansen	**3** b4
Hansteen	**11** e5

Harbinger Mts.

Harbinger Mts.	**6** h1, **7** b8, **8** c2
Harding	**8** e5
Harpalus	**6** f5, **8** a6
Hase	**15** d1, **16** c8
Hausen	**12** e3
Hecataeus	**15** b3
Heinsius	**10** e6, **12** a5
Heis	**5** g8, **6** f2
Helicon	**6** d4
Hell	**9** b1, **10** c7, **13** h1, **14** h7
Helmholtz	**14** b2, **16** g1
Henry, Paul	**11** f3
Henry, Prosper	**11** f3
Heraclides, Cape	**6** f4, **8** a5
Heraclitus	**10** b4, **14** e4
Hercules	**2** a4, **4** g6
Hercynian Mts.	**7** g8, **8** h2
Herigonius	**9** g6, **11** b5
Hermann	**7** f2, **11** e8
Herodotus	**7** d8, **8** e1
Herschel	**13** g7
Herschel, Caroline	**6** f2, **8** a4
Herschel, J.	**6** e7
Hesiodus	**9** d2, **10** e8
Hesiodus Cleft	**9** e2
Hevelius	**7** g3, **11** f8
Hiemis, Mare	**7** h2, **11** h8
Hind	**13** e6
Hippalus	**9** f3, **11** b2
Hipparchus	**13** e7
Holden	**15** c4
Hommel	**14** b4, **16** h3
Hook	**4** e5
Horrebow	**6** e6
Horrocks	**13** e7
Hortensius	**5** g3, **7** a4
Huggins	**10** b6, **14** f6
Humboldt, William	**15** c1, **16** b8
Humboldtianum, Mare	**4** f7
Humorum, Mare	**9** g4, **11** c3
Huygens, Mt.	**5** b6
Hyginus	**1** f3
Hyginus Cleft	**1** f3
Hypatia	**3** h1, **13** b7, **15** h7

I

Ideler	**14** c5
Imbrium, Mare	**2** h3, **5** e7, **6** d3, **8** a3
Inghirami	**12** f5
Iridum, Sinus	**6** e4, **8** a5
Isidorus	**15** f6

J

Jacobi	**10** a3, **14** e3
Jansen	**1** b3, **3** g4
Janssen	**14** a5, **16** f5
Julius Caesar	**1** d3
Jura Mts.	**6** e5. **8** a6

K

Kaiser	**10** a7, **14** e7
Kane	**2** d6
Kant	**13** b5
Kästner	**15** a6
Kelvin, Cape	**9** g3, **11** b2
Kepler	**7** c5
Kies	**9** e2, **10** f8, **12** a8
Kinau	**10** a3, **14** d3

Reimarus	16 e4	Sinas	1 a2, 3 f3	**U**	
Reiner	7 e4	Sirsalis	11 f5	Ukert	1 g3
Reinhold	5 f2, 9 d8	Sirsalis Cleft	11 f5	Ulugh Beigh	8 g3
Repsold	8 d6	Smythii, Mare	15 a7	Undarum, Mare	3 b2
Rhaeticus	1 f1, 13 e8	Snellius	15 d2, 16 c8	Ural Mts.	5 g1, 7 a2, 9 e7, 11 a6
Rheita	16 e6	Sömmering	5 c2, 9 a7, 13 h8		
Rheita Valley	16 e6	Somnii, Palus	3 d4		
Riccioli	7 h2, 11 g8	Somniorum, Lacus	1 b8, 2 a3, 4 g4	**V**	
Riccius	14 b7, 16 h6	Sosigenes	1 d3	Vaporum, Mare	1 f4
Riphaen Mts.	7 a1, 9 e6, 11 a5	South	6 f6, 8 b7	Vasco da Gama	7 h6
Ritchey	13 e5	Spallanzani	14 b5, 16 h4	Vega	16 d5
Ritter	1 c1, 3 h2, 13 b8	Spitzbergen Mts.	1 h8, 2 g2, 6 b2	Vendelinus	15 c4
Robinson	6 f6, 8 a7	Spörer	13 g7	Veris, Mare	11 h5
Rocca	11 g5	Spumans, Mare	3 b1, 15 b8	Vieta	11 f2, 12 f8
Römer	1 a6, 3 f7, 4 f2	Stadius	5 d4	Vitello	9 g2, 11 c1, 12 d8
Rook Mts.	11 h2	Stag's Horn Mts.*	9 b2, 13 h3	Vitruvius	1 a4, 3 f5
Roris, Sinus	6 g5, 8 c6	Steinheil	16 f4	Vlacq	14 a4, 16 g3
Rosenberger	14 a4, 16 g3	Stevinus	15 e1, 16 d7	Vögel	13 e4
Ross	1 c3, 3 h4	Stiborius	14 a7, 15 h1, 16 g7		
Rosse	15 f4	Stöfler	10 a6, 14 e6		
Rost	10 f3, 12 c3	Strabo	2 b7, 14 h8	**W**	
Rothmann	13 b1, 14 a8, 15 h2, 16 g7	Straight Range	6 d5	Wallace	5 c6
Rümker	8 d5	Straight Wall	9 b3, 13 h3	Walter	10 b7, 13 g1, 14 f7
Rutherford	10 d3, 14 g3	Street	10 d5, 14 h5	Wargentin	12 e5
		Struve	4 e5	Watt	16 f4
S		Struve, Otto	7 g7, 8 h1	Webb	3 b1, 15 b8
Sabine	1 c1, 3 h2, 13 b8	Suess	7 d4	Weigel	10 g3, 12 c3
Sacrobosco	13 c2	Sulpicus Gallus	1 e5	Weinek	15 f2, 16 f8
Santbech	15 e3	Sven Hedin*	7 h3	Weiss	9 d1, 10 f7, 12 a7
Sasserides	10 d6, 14 h6			Werner	10 a8, 13 f2, 14 f8
Saunder	13 d7			Whewell	1 e2, 13 c8
Saussure	10 c5, 14 g5			Wichmann	7 c1, 9 g7, 11 c6
Scheiner	10 e3, 12 b2	**T**		Wilhelm	10 e5
Schiaparelli	7 e7, 8 f1	Tacitus	13 c4	Wilkins	13 c1, 14 c8
Schickard	12 e5	Tannerus	14 c3	Williams	2 a3, 4 g5
Schiller	10 g4, 12 c4	Taquet	1 c4, 3 h5	Wilson	10 f2, 12 b1
Schmidt	1 d1, 13 d8	Taruntius	3 d2	Wöhler	14 a7, 16 g6
Schneckenberg, Mt.	1 f3	Taurus Mts.	3 e7, 4 e2	Wolf, Mt.	5 b5
Schomberger	14 d1	Taylor	13 c6	Wolf, Max	9 d3
Schröter	5 c2, 9 a8, 13 h8	Tempel	1 e2, 13 d8	Wollaston	8 d3
Schröter's Valley	7 d8, 8 e2	Teneriffe Mts.	2 h4, 6 c5	Wrottesley	15 d3
Schubert	3 a1, 15 a8	Thales	2 b7, 4 h8	Wurzelbauer	9 d1, 10 e7
Schumacher	4 e5	Theaetetus	1 f8, 2 e2		
Schwabe	2 c7	Thebit	9 a2, 13 g3		
Scoresby	2 f8	Theon Jnr.	13 c7	**X**	
Scott*	14 d1	Theon Snr.	1 d1, 13 c7	Xenophanes	8 c7
Secchi	3 d2	Theophilus	13 a5, 15 g6		
Seeliger	1 g1, 13 f7	Timaeus	2 f6, 6 a7	**Y**	
Segner	10 g3, 12 d3	Timocharis	5 d7, 6 c1	Yerkes	3 c4
Seleucus	7 f7, 8 g1	Timoleon	3 b7, 4 b4	Young	16 e6
Seneca	3 b6, 4 b2	Tisserand	3 d6, 4 d1		
Serao, Mt.†	5b 5	Torricelli	3 g1, 15 g7	**Z**	
Serenitatis, Mare	1 c6, 2 c1, 3 h7, 4 h2	Tralles	3 d7, 4 d2	Zach	10 b3, 14 e3
Sharp	6 g4, 8 b6	Tranquillitatis, Mare	1 b2, 13 a8, 15 g8	Zagut	13 c1, 14 b8, 16 h7
Shuckburgh	4 e5	Triesnecker	1 g2	Zeno	4 d6
Silberschlag	1 e2	Trouvelot	2 f5	Zöllner	13 b6
Simpelius	10 b2, 14 e2	Turner	5 d1, 9 b7	Zuchius	10 g2, 12 d3
		Tycho	10 d5, 14 h5	Zupus	11 e4

†This object cannot be identified; it lies between Mts. Wolf and Ampere, and is probably a peak in the Apennines. Its exact position is not indicated on the map.